Contents

Unit 1   Properties

Section 1   Presentation

1. Look and read:

Here are some properties of materials which are important in engineering:

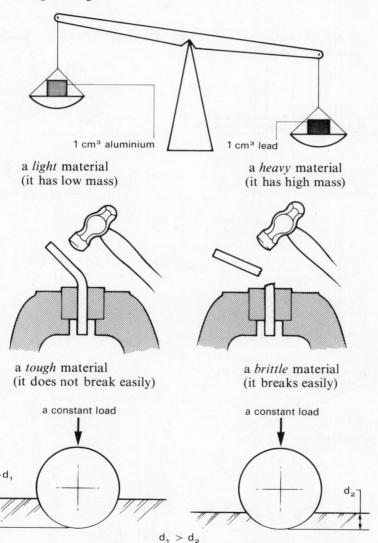

a *light* material
(it has low mass)

a *heavy* material
(it has high mass)

a *tough* material
(it does not break easily)

a *brittle* material
(it breaks easily)

a constant load

a constant load

$d_1 > d_2$

a *soft* material
(it can be dented or scratched easily)

a *hard* material
(it cannot be dented or scratched easily)

5

spring
an *elastic* material
(it returns to its original shape when
the deforming force is removed)

coin
a *plastic* material
(it does not return to its original
shape when the deforming force is
removed)

Plastic materials can be divided into two different groups, malleable
and ductile materials.

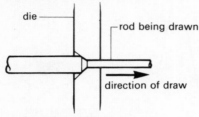

die

rod being drawn

direction of draw

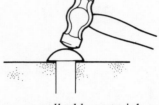

a *ductile* material
(it can be stretched)

a *malleable* material
(it can be hammered into shape)

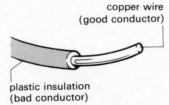

copper wire
(good conductor)

a good conductor of electricity
a bad conductor of electricity

plastic insulation
(bad conductor)

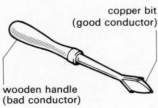
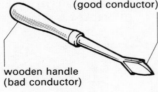

copper bit
(good conductor)

a good conductor of heat
a bad conductor of heat

wooden handle
(bad conductor)

unprotected
mild steel
corrodes
(goes rusty)
easily

polished
stainless
steel does
not go rusty

corrosion resistant
not corrosion resistant

Look at this table and the example:

In this table properties of materials are graded on a scale from A (maximum) to E (minimum): e.g. the property of heaviness is graded from very heavy (A) to very light (E).

Example: Copper is very heavy.
Nylon and rubber are very light.
Cast iron is heavy, but lighter than copper.
Glass is lighter than copper and cast iron, but heavier than aluminium.
Aluminium is light, but it is heavier than nylon and rubber.

	Copper	Aluminium	Cast iron	Glass	Nylon	Rubber
Heavy (A) — light (E)	A	D	B	C	E	E
Tough (A) — brittle (E)			D			
Hard (A) — soft (E)						E
A good conductor of electricity (A) — a bad conductor of electricity (E)		B				
A good conductor of heat (A) — a bad conductor of heat (E)						

Now complete the table with information from these sentences:

Copper is the toughest of the materials.
Aluminium, nylon and rubber are tough, but less tough than copper.
Cast iron is much more brittle than these materials but less brittle than glass.
Rubber is a very soft material.
Glass is a very hard material.
Cast iron is a hard material but it is not as hard as glass.
Nylon is harder than rubber.
Copper and aluminium are softer than cast iron and glass but harder than nylon and rubber.
Glass, nylon and rubber are very poor conductors of electricity and heat.
Copper, aluminium and cast iron are good conductors of electricity and heat, but copper is a better conductor than aluminium, and aluminium is a better conductor than cast iron.

2. Look at this sentence:

Glass is light compared with copper but heavy compared with nylon.

Now make similar true sentences from this table:

Copper Aluminium Cast iron Glass Nylon Rubber	is	light heavy brittle tough soft hard a good conductor of electricity a bad conductor of electricity a good conductor of heat a bad conductor of heat	compared with	copper aluminium cast iron glass nylon rubber

but	heavy light tough brittle hard soft a bad conductor of electricity a good conductor of electricity a bad conductor of heat a good conductor of heat	compared with	copper. aluminium. cast iron. glass. nylon. rubber.

3. Look at this table:

	Copper	Aluminium	Cast iron	Glass	Nylon	Rubber
Elastic (A) —— plastic (E)	D	D	D	B	C	A
Ductile (A) —— not ductile (E)	A	B	E	E	E	E
Malleable (A) —— not malleable (E)	A	A	D	E	E	E

Now say whether the following statements are true or false. Correct the false statements. Some statements refer to the table on page 7.

    a) Rubber is more elastic than nylon.
    b) Glass is not a malleable material.
    c) Copper is not very ductile.
    d) Aluminium is a ductile material but it is less ductile than copper.
    e) Copper is a very elastic material.
    f) Cast iron is very malleable.

g) Nylon is more malleable than cast iron.
h) Cast iron is heavier than copper.
i) Nylon is harder than rubber.
j) Cast iron is a better conductor of electricity and heat than aluminium.

4. Answer the following questions:

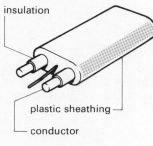

a) Why is copper used for electrical wire?

b) Why is plastic used to insulate electrical wire?

c) Why can mild steel be stretched into a tube?

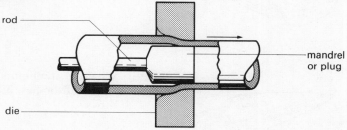

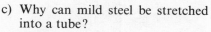

d) Why can silver (Ag) be hammered into shape?

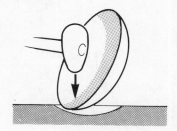

e) Why does steel need to be protected from moist air?

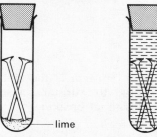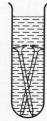

polished nails in dry air will not rust

polished nails in air free (boiled) water will not rust

polished nails in water containing air will rust

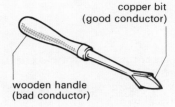

f) Why does a car spring keep its shape?

copper bit
(good conductor)

wooden handle
(bad conductor)

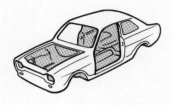

g) Why does this soldering iron have a wooden handle and a copper bit?

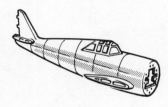

h) Why are car bodies made of steel sheets and not glass sheets?

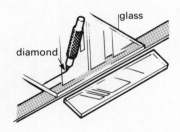

i) Why are aluminium alloys used in aircraft bodies, but not copper alloys?

glass

diamond

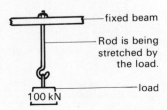

j) Why are diamonds used to cut glass?

Section 2   Development

5. Look and read:

fixed beam

Rod is being stretched by the load.

load

100 kN

Tensile strength is the ability of a material to withstand a stretching load without breaking.

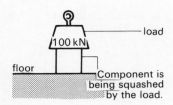

Compressive strength is the ability of a material to withstand a compressive load without breaking.

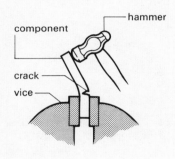

Rivet supporting a shearing load.

Shear strength is the ability of a material to withstand a shearing load.

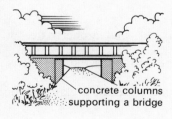

Impact strength (toughness) is the ability of a material to withstand an impact load.

Now complete these using the terms introduced above:

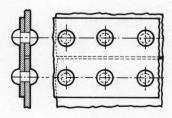

concrete columns supporting a bridge

a) Concrete is used for columns which support heavy loads because it has high _____ strength.

b) Rivets are made of metal which has high _____ _____.

c) The head of a hammer is made of
a metal which has high _____.

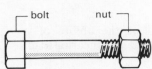

bolt   nut

d) Nuts and bolts are made of
metal which has high ____ ____.

6. Read the following definition:

A material which changes shape under load and returns to its original

shape $\begin{cases} possesses \\ has \\ exhibits \end{cases}$ the property of *elasticity*.

Now make sentences from this table:

A material	which changes shape	under a compressive load	possesses has exhibits	the property of	malleability. plasticity. ductility.
		under a tensile load			
	which exhibits plasticity	under load and keeps its new shape			

Now answer these questions:

a) What property does rubber possess?
b) When a metal bar is drawn into a rod, what property does it exhibit?
c) When the head of a rivet is hammered into shape, what property does it exhibit?
d) When gold (Au) is hammered, does it exhibit elasticity or plasticity?

7. Read these definitions:

Hardness is the ability to withstand scratching or indentation.
Corrosion resistance is the ability to resist chemical or electro-chemical attack.
A material which allows electricity to pass possesses *electrical conductivity*.
A material which melts easily has *high fusibility*.

A material which melts at a very high temperature has *low fusibility*.
A light material has *low density*. A heavy material has *high density*.

Now say whether the following statements are true or false. Correct the false statements.

    a) A compression spring possesses ductility.
    b) Aluminium has low resistance to corrosion by moist air.
    c) Metals with a low shear strength may be cut easily.
    d) Coins are not made with materials which possess elasticity.
    e) Solder has low fusibility.
    f) Plastics are used to insulate electrical wire because it has high thermal conductivity.
    g) Metals which are used in air frame parts have a low density.

8. What is the most important property needed by each of the following?

    a) a table leg   e) a railway line
    b) a car radiator   f) a chisel
    c) a fire brick   g) a battery terminal
    d) a rivet   h) the cables on a suspension bridge

Section 3   Reading

9. Read this passage:

Ferrous metals
Ferrous metals contain iron (Fe). One kind of ferrous metal is cast iron. It has high compressive strength, and is easy to cast, because it has high fluidity and relatively high fusibility. However, it has low malleability and a relatively low tensile strength.

Wrought iron has greater toughness than cast iron. It is very malleable and ductile. It has relatively high corrosion resistance. It is often used for the chains of anchors and cranes.

Mild steel is more often used in industry because it has a lower cost and a slightly higher strength than wrought iron. It has high malleability and ductility. It can be easily worked when it is cold and when it is hot. It has many uses, including the production of car bodies, rods and bars, nuts and bolts etc.

Now answer these questions:

    a) Which is the most important element which is common to all ferrous metals?
    b) Why are motor cycle cylinders made of cast iron?
    c) Why are chains not made of cast iron?
    d) Why are anchor chains made of wrought iron?
    e) Why is mild steel commonly used in industry?
    f) Why can mild steel be used for car bodies?
    g) Why is mild steel used for rods and bars?

Section 4 Listening

10. Read the following statements then listen to the passage and number the statements in the order in which you hear them:

    a) Lead is very heavy.
    b) Aluminium possesses high ductility.
    c) Zinc possesses resistance to atmospheric corrosion.
    d) Copper is a good conductor of electricity.
    e) Aluminium is very malleable.
    f) Lead cannot withstand a heavy tensile load.
    g) Zinc is a non-ferrous metal.
    h) Aluminium is a light metal.
    i) Zinc is soft.
    j) Lead possesses relatively high fusibility.

11. Answer these questions:

    a) Why is copper used for a soldering iron bit?
    b) Why does solder contain lead?
    c) Why are steel sheets sometimes covered with zinc?
    d) Which has higher tensile strength, aluminium or copper?
    e) Which is heavier, lead or zinc?
    f) Which has a higher melting point, lead or zinc?

Unit 2   Location

Section 1   Presentation

1. Look and read:

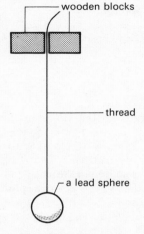

A simple pendulum

The wooden blocks are *at the top* of the pendulum.
The lead sphere is *at the bottom* of the pendulum.
The thread is *between* the wooden blocks.
The wooden blocks are *opposite* each other.
The wooden blocks are *on either side of* the thread.
The thread is *held in place by* the wooden blocks.
The wooden blocks are *over* the lead sphere.
The lead sphere is *under* the wooden blocks.
The lead sphere is *at the end* of the thread.

Now ask and answer questions like the following:

Where are the wooden blocks located?
At the top of the pendulum *or* on either side of the thread etc.

What is at the end of the thread?
A lead sphere.

2. Now look at this diagram:

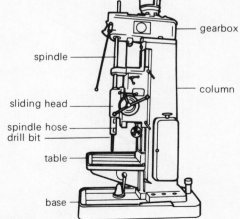

Column type drilling machine

15

Note the use of *in alignment with*.

Example: The drill bit must be *in alignment with* the centre of the hole in the workpiece.

Complete these sentences:

  a) The gearbox is ...... the drill.
  b) The base is ...... the drill.
  c) The table is ____ the drill and the drill bit.
  d) The drill bit must be ...... with the spindle.
  e) The drill bit is ...... by the spindle nose.
  f) The column is ____ the gearbox and the base.
  g) The spindle nose is ...... the spindle.

3. Read this:

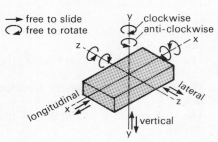

Six degrees of freedom

The body in the diagram can be moved *longitudinally* (back and forth), *laterally* (from side to side) and *in the vertical plane* (up and down). The body is thus *free to slide* along the X, Y, Z axes (plural of axis). The body is also *free to rotate clockwise* and *anti-clockwise about* the X, Y, Z axes.

When a body is worked upon, it is held in place by restraining its movement. The diagram below shows how these restraints are built up.

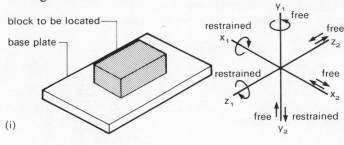

(i)

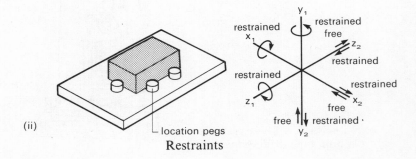

(ii)

Restraints

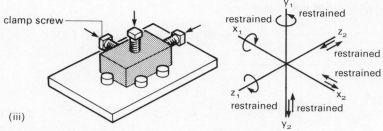

(iii)

Look at block (i) and make sentences from this table:

The block is	free to slide restrained partially restrained	longitudinally. laterally. in the vertical plane. along the X axis. along the Y axis. along the Z axis.
	free to rotate restrained	about the X axis. about the Y axis. about the Z axis.

Now look at block (ii) and say whether the following are true or false. Correct the false statements.

    a) The block is free to slide in the direction of Z_1.
    b) The block is free to slide in the direction of Z_2.
    c) The block is free to slide laterally in the direction of X_1, but not in the direction of X_2.
    d) The block is free to slide in the vertical plane both in the direction of Y_1 and in the direction of Y_2.
    e) The block is free to rotate about the X axis.
    f) The block is restrained about the Y axis.

4. Now answer these questions:

    a) In which of the three diagrams is the block free to slide longitudinally?
    b) In which of the diagrams is the block completely restrained along the X axis?
    c) In which of the three diagrams is the block free to move up but not down?
    d) In which of the diagrams is the block free to slide from side to side?
    e) In which of the diagrams is the block free to slide in the direction of Z_2 but not in the direction of Z_1?
    f) In which of the diagrams is the block free to rotate about the Y axis?
    g) In which of the diagrams is the block restrained in the direction of Z_2?
    h) In which of the diagrams is the block completely restrained?

17

5. Look at these lines:

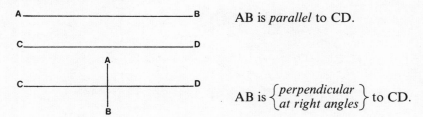

AB is *parallel* to CD.

AB is $\left\{\begin{array}{l}perpendicular \\ at\ right\ angles\end{array}\right\}$ to CD.

Now look at the diagram of the drilling machine again, and choose the correct alternative in these sentences:

> *Example:* The movement of the sliding head is parallel/perpendicular to the spindle axis.
> The movement of the sliding head is *parallel* to the spindle axis.

a) The lateral movement of the table is parallel/perpendicular to the spindle.
b) The vertical axis of the drill bit is parallel/perpendicular to the surface of the workpiece.
c) The horizontal axis of the sliding head is parallel/perpendicular to the spindle axis.
d) The horizontal axis of the gearbox is parallel/perpendicular to the horizontal axis of the base.
e) The drill bit is free to rotate in a clockwise/an anti-clockwise direction.
f) The drill bit moves up and down/from side to side.

6. Look at this diagram:

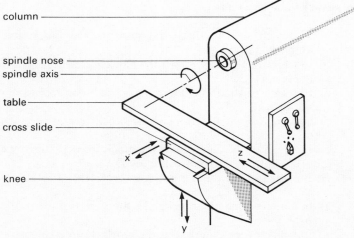

Movements on a milling machine

18

Name these parts of the milling machine:

a) This part slides longitudinally on the cross-slide.
b) This part slides laterally on the knee.
c) This part rotates in a clockwise direction.
d) This part is lying in a plane parallel to the table.
e) This part slides in a perpendicular plane in front of the machine column.
f) The movement of this part is perpendicular to the front of the machine column.
g) The working surface of this part is perpendicular to the front of the machine column.
h) This part is parallel to the table but perpendicular to the front of the machine column.

Section 2   Development

7. Look at these:

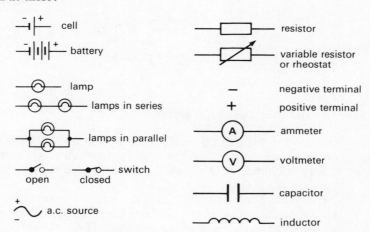

Electrical symbols

Look and read:

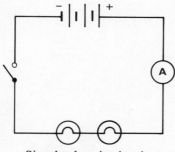

Simple electric circuit

In the diagram, the battery is positioned at the top of the circuit diagram. The switch is on the left and is open. The lamps are positioned at the bottom of the diagram, and are connected in series. The ammeter is on the right and is in series with the other components.

Now look at these circuits and describe them:

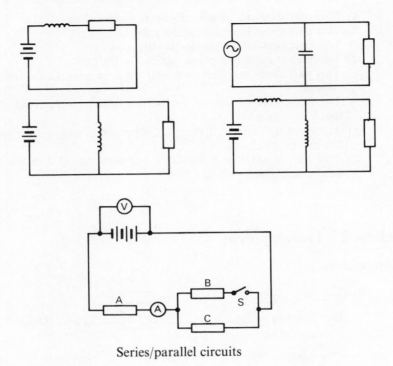

Series/parallel circuits

Now read this and draw the circuit described:

At the top of the circuit there is a battery with three cells. The positive terminal is on the right and the negative terminal is on the left. This battery is connected in parallel with a voltmeter. This should be shown in the diagram as above the battery.

The sides have nothing connected into the circuit but at the bottom of the circuit there are three lamps connected in series. In parallel with each lamp is a voltmeter. These should be shown under each of the lamps. There is a switch connected in series with the lamps. This switch is closed.

Section 3   Reading

8. Read this:

When a hole is drilled in a component, the machinery and the workpiece are located as follows:

- i) The axis of the drill must be in alignment with the axis of the drilling machine spindle.
- ii) The drill must be restrained by the spindle.
- iii) The centre lines of the holes in the workpiece must be in alignment with the axis of the spindle.
- iv) The workpiece must be restrained.

Now write sentences to explain what is wrong with each of these diagrams:

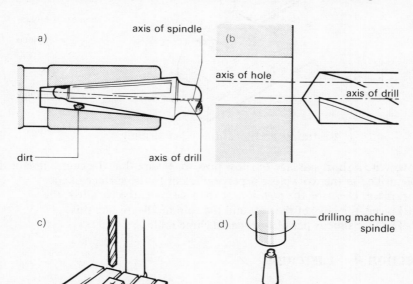

Drilling faults

9. Look and read:

This diagram shows the alignments for drilling a hole perpendicular to the workpiece. To do this, it is necessary to ensure that the spindle axis is perpendicular to the workpiece and the following alignments must be checked:

i) A pair of parallels, of exactly the same size, must be placed under the workpiece.

ii) The working surface of the vice slide must be parallel to the machine table.

iii) The fixed jaw of the vice must be perpendicular to the machine table.

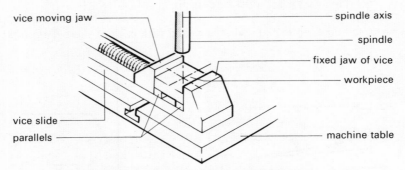

Restraints and locations when drilling

Now write a short passage to show how we ensure that the centre line of the hole drilled in the workpiece is perpendicular to the surface of the workpiece. Describe the position of the vice slide, the parallels, the workpiece, the jaws of the vice and the spindle. Begin like this:

The vice slide is parallel to the machine table . . .

Section 4   Listening

10. Listen to the description of eight circuits. Draw the circuits as you listen.

Unit 3   Structure

Section 1   Presentation

1. Look at this table:

Composition of alloys (the ticks indicate the metal is present in the alloy)

Name of alloy	Copper (Cu)	Zinc (Zn)	Tin (Sn)	Other metals
Standard brass	✓	✓		
Free-cutting brass	✓	✓		Lead (Pb)
Naval brass	✓	✓	✓	
High tensile brass	✓	✓	✓	Manganese (Mn) Iron (Fe) Aluminium (Al)
Phosphor bronze	✓		✓	Phosphorous (P)
Admiralty gunmetal	✓	✓	✓	

Make questions and answers like the following:

Question: What is standard brass *composed of*?
Answer: It is *composed of* copper and zinc.
Question: What are the *constituents* of standard brass?
Answer: The *constituents* of standard brass are copper and zinc.
Question: Does standard brass *contain* zinc?
Answer: Yes, it does.
Question: Does standard brass *contain* tin?
Answer: No, it doesn't.
Question: Which alloy *contains* lead?
Answer: Free cutting brass.

2. Look at this diagram:

Engineering materials

Metals
- Ferrous metals (These contain iron.)
- Non-ferrous metals and alloys (These do not contain iron.)

Non-metals
- Synthetic materials (plastics etc.)
- Natural materials (rubber, glass, wood, ceramics, diamonds, oil.)

Now read this:

Engineering materials *are divided into* two groups, metals and non-metals. Metallic materials *consist of* ferrous metals, which *contain* iron and non-ferrous metals, which *do not contain* iron. Non-metals *are divided into* two groups, synthetic materials and natural materials. Synthetic materials *include* plastics while natural materials *include* rubber, glass, wood, diamonds, oil etc.

23

Look at this diagram:

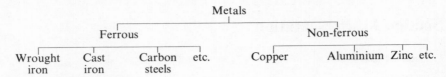

Now complete this:

> Metals ...... two groups, ferrous and non-ferrous. Ferrous metals
> ____ iron, while non-ferrous .... Ferrous metals —— wrought
> iron, cast iron, carbon steels etc., while non-ferrous metals include
> ....

3. Read this:

> Steels are divided into two groups, plain carbon steels and alloy
> steels. Plain carbon steels consist of alloys of iron and carbon which
> contain up to 1·5% carbon and up to 1% manganese. Alloy steels
> consist of steels which contain more than 1% manganese and other
> metals. Alloy steels include high tensile steels, heat resisting steels,
> and stainless steels. Carbon steels include dead mild steel, which
> contains from 0·1% to 0·15% carbon; mild steel, which contains
> from 0·15% to 0·25% carbon; medium carbon steel, which contains
> from 0·25% to 0·85% carbon and high carbon steel, which contains
> from 0·85% to 1·5% carbon.

Now complete the diagram below:

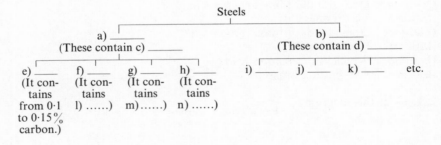

4. Say which group the following steels belong to:

> *Example:* a 0·15% carbon steel for forming motor car body panels
> It belongs to the dead mild steel group.

a) a 0·6% carbon steel for making chisels
b) a 1·2% plain carbon steel wire for making coil springs

c) a 0·25% plain carbon steel for machining nuts and bolts
d) a steel suitable for the cables in a suspension bridge
e) a steel suitable for the valves in a motor car engine
f) an alloy steel used for storing drinking water

Section 2 Development

5. Look at this diagram:

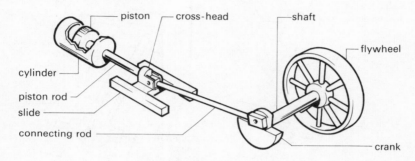

Steam engine

Look at the left hand side of the diagram and read this:

> The piston is *fixed to* the piston rod.
>   i.e. the piston is not free to move in relation to the piston rod.
> The connecting rod is *attached to* the cross-head.
>   i.e. the connecting rod is free to move in relation to the cross-head.
> The piston is *connected/joined to* the cross-head *by* the piston rod.
>   i.e. they are joined together by an intermediate member.
> The part which contains the piston *is called* the cylinder.
> The piston is *in contact with* the walls of the cylinder.
> The cross-head is *mounted on* a slide.
>   i.e. it is partially restrained by the slide, but it is not fixed to it.
> These parts are *made of* metal.

Now look at the right hand side of the diagram and complete these sentences:

a) The part which maintains momentum ...... the flywheel.
b) The flywheel is ...... a shaft.
c) The crank is ...... the flywheel by a shaft.
d) The connecting rod is ...... the crank.
e) The cross-head is ...... the crank ....
f) These parts ...... metal.
g) The engine must be ...... supports.

6. Look and read:

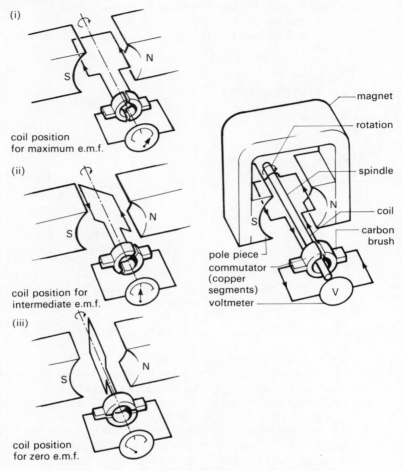

Principle of the d.c. generator

The principle of the direct current (d.c.) generator is that a coil is rotated in a magnetic field, and a current is thus induced in the coil. The current is picked up by the carbon brushes, which are in contact with the commutator.

Now describe the locations and structure of the various parts using the following notes:

    a) The generator/consists/magnet, coil etc.
    b) The coil/between/pole pieces/mounted/spindle. Fixed/commutator
    c) The pole pieces/fixed/either side/magnet
    d) The commutator/made/copper. In contact with/brushes

e) The brushes/made/carbon. Connected/external circuit
f) When/the coil/moving/perpendicular/magnetic field, the e.m.f./at its maximum
g) When/ the coil/moving at an angle/magnetic field, the e.m.f./ intermediate
h) When/the coil/moving/parallel/magnetic field, the e.m.f./nil

Section 3   Reading

7. Read this:

Pure metal is composed of three-dimensional shapes called crystals. All solid metals are crystalline and the crystal structure determines the properties of the metal and its use. A metal which has a coarse grain structure usually has low tensile strength, and is soft and ductile and can therefore be cold-worked. Material which has a fine grain structure is strong, tough and less workable.

Each crystal contains atoms which are arranged in fixed frameworks or networks. These are called space lattices.

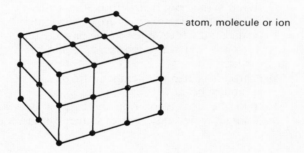

— atom, molecule or ion

A space lattice

In three-dimensional crystals there are fourteen different kinds of space lattice, but the atoms of nearly all metals are arranged on only three of those space lattices. These are face-centred cubic, body-centred cubic and close-packed hexagonal.

close-packed hexagonal   face-centred cubic   body-centred cubic

Typical metallic crystal structures

27

At room temperature the face-centred cubic structures are found in the ductile pure metals such as copper, lead, nickel, gold, silver, platinum and aluminium. The body-centred cubic structures are found in pure metals, which may be either ductile or brittle, such as chromium, tungsten, iron, molybdenum and vanadium. Examples of metals with close-packed hexagonal structures include zinc, titanium and magnesium.

Some metals, principally iron, manganese, tungsten and cobalt have more than one possible crystal structure. Iron is body-centred cubic from room temperature to about 900°C. Above 900°C it becomes face-centred cubic and stays in this condition up to 1,400°C when it again becomes body-centred cubic. Metals which change in this way are said to be allotropic.

Now answer these questions:

    a) Does an annealed mild steel suitable for cold working contain large or small grains?

    b) Does a steel suitable for a motor car body panel have a fine or coarse grain structure?

    c) Why isn't molten iron granular?

    d) Which kind of crystal structure contains one atom in the centre?

    e) Which kind of crystal structure contains three atoms in the centre?

    f) What is the crystal structure of zinc?

    g) What is the crystal structure of chromium?

    h) Is nickel ductile? What in the passage tells you this?

    i) If iron is heated to a temperature of 1,000°C, what will its crystal structure be?

    j) If iron is heated to a temperature of 1,450°C, what will its crystal structure be?

    k) Will the density of iron increase or decrease slightly when it changes from body-centred cubic to face-centred cubic?

Section 4   Listening

8. Copy this diagram. Then listen to the passage, complete the diagram, and add these labels.

glass container
gas
filament
terminals
base
lugs

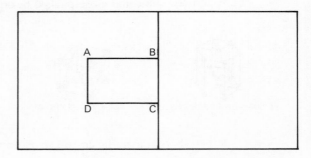

Unit A Revision

1. Look and read:

This building consists of a steel framework and a covering. The framework is composed of vertical and horizontal inclined members.

The vertical members, which are called columns, are located in concrete bases. The roof structure is supported by the columns which are joined longitudinally by the wall plates. These are attached to the tops of the columns. The bottoms of the rafters are fixed to the wall plates and the rafter tops are fitted to the ridge pole. The horizontal, longitudinal members which connect the rafters are called the purlins.

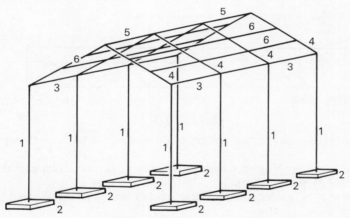

Frame of a pre-fabricated building

Now match the numbers with the names of the parts.

2. Look at this diagram:

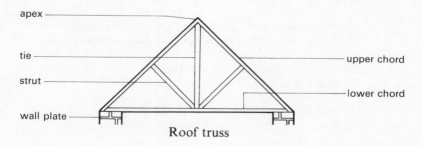

Roof truss

29

Read this:

> The horizontal member which is located at the base of the truss is called the lower chord. It is attached to the bottom end of each of the upper chords. It is fixed to and supported by the wall plates.

Now write similar descriptions of the upper chords, the struts and the tie of the roof truss. Begin your description of the upper chords like this:

> The inclined members which are located between the ends of the lower chord and the apex . . .

3. Look at these diagrams:

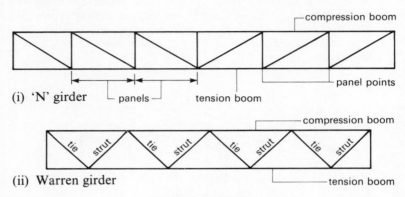

(i) 'N' girder

(ii) Warren girder

Types of girder

Compare and contrast the two girders using sentences like these examples:

> In *both* the 'N' girder *and* the Warren girder the struts and the ties meet at panel points.
> In the 'N' girder there are five struts *whereas* in the Warren girder there are six struts.

Use the following information:

> The compression boom is at the top of the girders.
> The tension boom is at the bottom of the girders.
>
> The struts in the 'N' girder are vertical members.
> In the Warren girder they are inclined members.
>
> The panels in the 'N' girder are rectangular in shape.
> The panels in the Warren girder are triangular in shape.
>
> There are six panels in the 'N' girder.
> There are seven panels in the Warren girder.
>
> The ties are inclined members in the 'N' girder.
> They are also inclined members in the Warren girder.

30

4. Read this passage:

Mild steel is a very important building material widely used in structural engineering. Its properties are as follows: it can exhibit the property of elasticity, and is ductile, malleable, weldable, and fairly tough and soft. It contains up to 0.25% carbon and its breaking strength in compression and tension is 430 to 510 N/mm^2.

Because of these properties it is a very useful structural material from which beams and other structural sections can be manufactured by hot rolling. These sections are used in many ways such as (i) struts, which must withstand compressive stresses; (ii) ties, which must withstand tensile stresses; (iii) columns, which must withstand bending and compressive stresses; (iv) chords and booms, which must withstand a combination of tensile, compressive and shear stresses; (v) reinforcement for concrete.

Now answer these questions:

a) Can mild steel be drawn into wire?
b) Can mild steel be hammered into shape?
c) Can mild steel be welded?
d) Is the strength of mild steel greater in compression than in tension?
e) Which members mentioned in the passage must withstand shear stresses?
f) Which members mentioned in the passage must withstand a combination of stresses?
g) Why is steel used in concrete beams?
h) Are heavy structural steel sections normally hot or cold-rolled?

Unit 4 Measurement 1

Section 1 Presentation

1. Look at this table:

Name	Chemical symbol	Density kg/m³ (kilograms per metre cubed)	Melting point 0°C (degrees Celsius)	Average tensile strength N/mm² (Newtons per millimetre squared)
Aluminium	Al	2,700	660	90
Magnesium	Mg	1,700	651	248
Copper	Cu	8,950	1,083	340
Cast iron	Fe	7,860	1,535	116
Lead	Pb	11,300	327	15
Nickel	Ni	8,900	1,458	172
Titanium	Ti	4,500	1,725	216
Zinc	Zn	7,100	420	110

Look at these sentences:

Aluminium *has a density of 2,700 kg/m³* (two thousand seven hundred kilograms per metre cubed).
Lead *has a tensile strength of 15N/mm²* (fifteen Newtons per millimetre squared).
Copper *has a melting point of 1,083°C* (one thousand and eighty-three degrees Celsius).

Now make similar statements about other metals.

2. Look at these sentences which show how the densities may be compared for the various metals:

Magnesium has a *very low* density.
Aluminium has a *low* density.
Titanium has a *fairly low* density.
Cast iron has a *fairly high* density.
Lead has a *very high* density.

Now name these metals:

a) A metal with a fairly high density, a very high melting point and a fairly low tensile strength.
b) A metal with a fairly high density, high tensile strength, and a very high melting point.
c) A metal with a low density, a fairly low melting point and low tensile strength.

32

d) A metal with a very low melting point, fairly low tensile strength and a fairly high density.

e) A metal with very low density, a fairly low melting point and very high tensile strength.

f) A metal with high density, a fairly high melting point and very high tensile strength.

g) A metal with high tensile strength, a high melting point and a high density.

h) A metal with a very low melting point, very high density and very low tensile strength.

3. Look at these examples:

The *lowest* melting point of soft solders is 183°C.
The *highest* melting point of soft solders is 310°C.
So the melting points of soft solders *range from* 183°C *to* 310°C.
The melting points of soft solders *have a range of* 127°C.

Now complete these sentences:

a) The lowest melting point of silver solders is 620°C.
The highest melting point is 740°C.
The melting point of silver solders ranges from ____ to ____ .

b) The ____ melting point of brasses is 850°C.
The ____ melting point of brasses is 1,000°C.
The melting points of brasses ....

c) The lowest ____ of Monel metal is 1,300°C.
It has a range of 150°C.
Therefore the highest ____ .

Section 2   Development

4. Look and read:

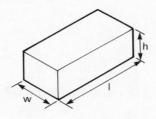

Volume $= w \times l \times h$
Area (of side) $= h \times l$
Area (cross-sectional) $= w \times h$
Surface area $=$ total area of all surfaces

Area and volume of rectangular prisms

Now look at this diagram and complete the measurements. Then make statements like the following:

The rectangular prism has a length of 100 mm (one hundred milli-metres).

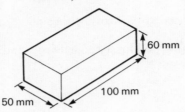

Length 100 mm
Height ......
Width ......
Cross-sectional area ......
Surface area ......
Volume ......

60 mm
100 mm
50 mm

Look at these diagrams:

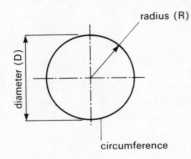

radius (R)

diameter (D)

circumference

$\text{Area} = \pi R^2 = \pi D^2/4$
$\text{Circumference} = \pi D = 2\pi R$

$\text{Cross-sectional area} = \dfrac{\pi}{4}(D^2 - d^2)$

$\text{Volume} = \text{cross-sectional area} \times \text{length}$

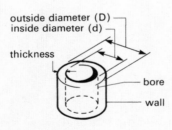

outside diameter (D)
inside diameter (d)
thickness
bore
wall

Area and volume of cylinders

Now complete these measurements. Then make statements about the *outside diameter, inside diameter*, etc.

30 mm
100 mm
20 mm

Outside diameter ......
Inside diameter ......
Length ......
Cross-sectional area ......
Surface area of bore
Volume ......

5. Read this:

To calculate the potential energy of a body
The *formula* for calculating potential energy is:
Potential energy = mgh
   i.e. Potential energy equals m times g times h.
Where m = mass in kilograms
     g = acceleration due to gravity $9 \cdot 8$ m/sec^2
            (nine point eight metres per second squared)
 and h = height in metres

Problem: A 10 kilogram ball is raised through a height of 10 metres. What is its potential energy in joules?

Complete this:

Solution: The ball has a mass of ____ .
The height through which the ball is raised is ____ .
Therefore the potential energy of the ball is ____ times ____ times ____ equals ____ joules. (Expressed numerically ____ × ____ × ____ = ____ J.)

Now look at the following problem and formula and make complete statements about the formula, calculation and solution as in the example of potential energy above:

Formula: Kinetic energy = $\frac{1}{2}$ mv^2
       where m = mass in kilograms
              v = velocity in metres per second

Problem: Using this formula calculate the kinetic energy of a 10 kg ball which is released and travels 5 metres in a straight path for one second.

6. Read this:

A machine spindle has different speeds which the operator selects according to the work. These speeds are usually arranged in a geometric progression. For example spindle speed no. 1 could be 40 revs/min and the constant multiplier could be $1 \cdot 5$.

Look at these examples:

The constant multiplier is $1 \cdot 5$.
Spindle speed no. 1 is 40 rev/min (forty revs per minute).
Spindle speed no. 2 is 60 rev/min (i.e. 40 times $1 \cdot 5$).
Spindle speed no. 3 is 90 rev/min (i.e. 60 times $1 \cdot 5$).

Now complete the following:

a) Spindle speed no. 4 is ＿＿ rev/min (......).
b) Spindle speed no. 5 is ....
c) Spindle speed no. 6 is ....
d) Spindle speed no. 7 is ....
e) Spindle speed no. 8 is ....

Using these axes copy and complete the graph:

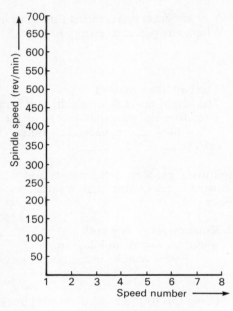

Spindle speeds

7. Look at these sentences:

Spindle speed no. 2 is *exactly* one and a half times *as* $\left\{\begin{matrix} great \\ fast \end{matrix}\right\}$ *as* speed no. 1.

Spindle speed no. 5 is *approximately* five times *as* $\left\{\begin{matrix} great \\ fast \end{matrix}\right\}$ *as* speed no. 1.

Now compare the following:

a) Spindle speeds no. 3 and no. 1
b) Spindle speeds no. 4 and no. 1
c) Spindle speeds no. 6 and no. 2
d) Spindle speeds no. 7 and no. 3
e) Spindle speeds no. 6 and no. 5
f) Spindle speeds no. 7 and no. 5

Section 3   Reading

8. Read this:

Most physical quantities can be described completely by a statement of magnitude followed by a unit of measurement. Examples include time, price, temperature, area, volume and speed. We say 10 minutes, 50 dollars, 37°C, 200 metres squared, 6 litres and 50 kilometres per hour. All of these statements about a quantity consist of a number giving the amount, followed by a unit of measurement. Quantities which can be described in this way are known as scalar quantities, or scalars.

Other quantities, however, cannot be completely described in this way. Vector quantities, or vectors, require a statement of direction in addition to a statement of amount, and a unit of measurement. Examples of vectors include force, velocity, acceleration and displacement.

Let us consider two examples of the difference between a scalar quantity and a vector quantity. First of all let us take speed and velocity. The speed of a body is the distance travelled divided by the time taken for a given journey. The standard unit of time is the second, the standard unit of distance is the metre and, therefore, speed can be expressed in metres per second. Speed, then, is a scalar quantity since direction is not specified. The velocity of a body, however, is a vector quantity because it includes not only its speed but also the direction in which the body travels. Secondly we have mass and weight. The mass of a body is the quantity of matter which the body contains. The standard unit of mass is the kilogram, but very small masses are measured in grams. The weight of a body, however, is a vector quantity, because it includes both the mass and the force which gravity exerts on the body. This force must have a direction. Weight is measured in newtons. There is a simple relationship between the mass of a body and its weight, namely:

Weight in newtons = 9·81 (acceleration due to gravity) × mass in kilograms.

Now answer these questions:

a) What is the difference between a vector quantity and a scalar quantity?
b) Why is velocity a vector quantity?
c) A car travels at 80 km/h. Is this a statement of speed or velocity?
d) In statement (c) what is the magnitude, and what is the unit of measurement?
e) If a car travels in a circle at 80 km/h what changes – its speed or its velocity?
f) What is the difference between mass and weight?
g) A workpiece has a mass of 10 kg. What is the weight of this workpiece?

37

h) Look at this diagram and answer the questions:

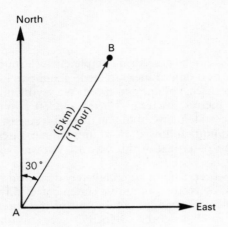

Speed and velocity

What is the distance between A and B?
If a man walks from A to B in one hour, what is his speed?
What is his velocity?

9. Listen to the passage and carry out the instructions.

Unit 5　Process 1 Function and Ability

Section 1　Presentation

1. Look at this table:

The properties and uses of metals

Name	Properties	Some uses
Aluminium	Low density. Soft and ductile. High electrical and thermal conductivity.	Lightweight electrical conductors.
Copper	Ductile and malleable. Medium tensile strength. High electrical and thermal conductivity. High heat capacity. Corrosion resistant.	Electrical conductors and heat exchangers.
Lead	Soft and ductile. Low tensile strength. Corrosion resistant.	Electric cable sheaths.
Tin	Corrosion resistant.	Coating sheet mild steel to give tin plate.
Zinc	Soft and ductile. Corrosion resistant.	Coating sheet steel to give galvanised steel.
Wrought iron	Ductile and malleable. Fairly high tensile strength.	Chains. Crane hooks.
Medium carbon steel (hardened and tempered)	Tough and relatively hard.	Cold chisels. Motor car crankshafts. Springs.

Ask and answer questions like the following:

Why is aluminium *used for* lightweight electrical conductors?
Because it has high electrical conductivity and low density.

2. Look at this sentence:

Aluminium has high electrical conductivity and low density and is, therefore, used for lightweight electrical conductors.

Now write six similar sentences.

3. Look at these sentences:

A hard material $\begin{Bmatrix} \textit{has the ability to} \text{ withstand} \\ \textit{is able to} \text{ withstand} \\ \textit{is capable of} \text{ withstanding} \end{Bmatrix}$ scratching or indentation

by another hard body.

Now say whether these statements are true or false. Correct the false statements.

a) Aluminium is capable of conducting electricity.
b) Lead is unable to resist attack by corrosive substances.
c) Wrought iron is not capable of withstanding a tensile force without breaking.
d) Copper is able to withstand hammering.
e) Tin does not have the ability to withstand corrosion.
f) Medium carbon steel is capable of withstanding scratching or indentation.
g) Aluminium does not have the ability to conduct heat.
h) Copper is capable of being drawn into wire.
i) Lead is unable to withstand scratching from another body.

Section 2　Development

4. Look and read:

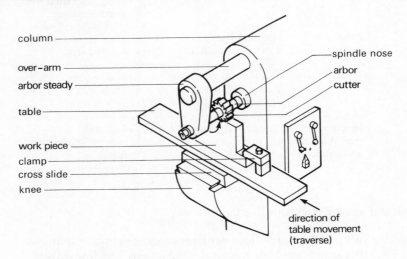

Movements on a milling machine

The spindle nose *acts as* a location for the arbor (the shaft which carries the cutters).
The function of the table is *to* support the workpiece. It also *enables* the workpiece to move against the cutter.

The cross-slide *provides* the operator with a means of centring the workpiece under the cutter.

The knee *enables* the workpiece to be fed into the cutter, thus *controlling* the depth of metal that is removed.

Answer these questions:

a) Which part acts as a support for the workpiece?
b) What does the cross slide enable the operator to do?
c) Which part locates the arbor?
d) Which part provides a means of controlling the amount of metal that is removed?
e) What is the function of the knee?
f) What is the function of the spindle nose?

5. Look at this diagram:

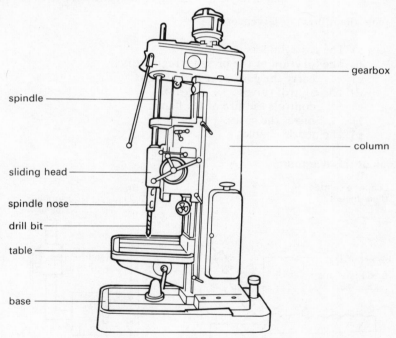

Column-type drilling machine

Now make true sentences from this table:

The function of the	base chuck spindle sliding head gearbox	is to	house the gears. transmit the drive to the drill. hold the machine rigid. hold the drill in place. control the vertical movement of the drill.

6. Look at this diagram:

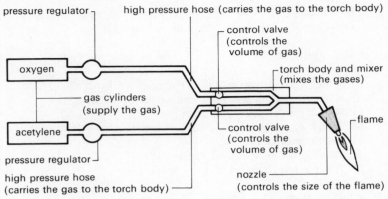

Oxy-acetylene welding equipment

Complete the following sentences:

    a) The gas cylinders ....
    b) The function of the pressure regulators ....
    c) ...... carry the gas to the torch body.
    d) The control valves ....
    e) ...... controls the size of the flame.
    f) ...... mixes the gases.
    g) The nozzle enables ....

7. Look at this diagram:

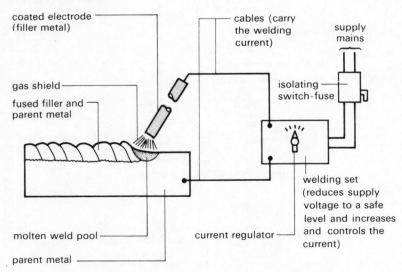

Electric arc welding equipment

Complete the following using these words:

> act(s) as, function, provides, enables, connected

In the electric arc welding equipment the workpiece and the electrode _____ terminals in the electric circuit, and are _____ by cables to a welding set. This welding set _____ the electricity. The main _____ of the electrode is to carry the current which melts the surface of the workpiece, but it also _____ a filler metal. One _____ of the coating on the electrode is to protect the weld against impurities, another is to _____ an insulator for the wire. In addition, it _____ a gas shield around the molten metal to keep oxygen and nitrogen in the air away from the metal, and _____ alternating current to be used.

Section 3   Reading

8. Read this:

> A relay is a switching device for opening and closing one or more electrical circuits on receipt of an electrical signal.
>
> A simple kind of relay consists of two parts, a switch in a high voltage circuit which is operated by a spring loaded armature and a low voltage electro-magnetic circuit.
>
> The electro-magnet comprises a soft iron core shaped like a horseshoe. Each arm of the core has a coil wound around it. The coils are connected to each other and to a battery through an on/off switch.
>
> When the switch is closed, the electro-magnet is energised, and the armature is attracted towards it. This closes the contacts and enables the current to flow through the high voltage circuit. When the switch is opened, a spring pulls the armature away from the magnet, and the contacts open again.

Now match the letters with the parts of a relay:

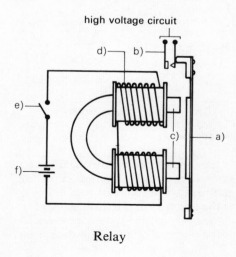

Relay

9. Say whether the following statements are true or false. Correct the false statements.

a) When the switch in the low voltage circuit is closed, the high voltage circuit is broken.
b) When the contacts in the high voltage circuit are closed, current flows from the low voltage circuit to the high voltage circuit.
c) When the switch in the low voltage circuit is open, the armature moves towards the magnet.
d) The contacts in the high voltage circuit are controlled by the electro-magnet.
e) The function of the soft iron core is to open and close the contacts in the high voltage circuit.
f) The armature is restrained.
g) The electro-magnet is attached to the battery.
h) The coils of wire are mounted on the soft iron core.
i) The battery and switch are connected in parallel.
j) When the contacts are closed, they act as conductors.

Section 4   Listening

10. Listen to the passage and make notes under these headings:

a) The function of a centrifugal governor.
b) The construction of a centrifugal governor.
c) What happens to the spherical masses when the shaft rotates.
d) The function of the throttle valve.
e) What happens when the engine speed decreases.

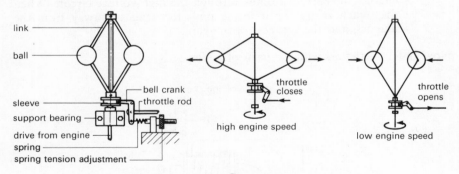

Governor

11. Now answer these questions:

a) What is the function of a centrifugal governor?
b) What does a centrifugal governor consist of?
c) What happens when the shaft rotates?
d) What is the function of the throttle valve?
e) What happens when the speed decreases?
f) What is the purpose of adjusting the spring?

Unit 6   Process 2 Action in Sequence

Section 1   Presentation

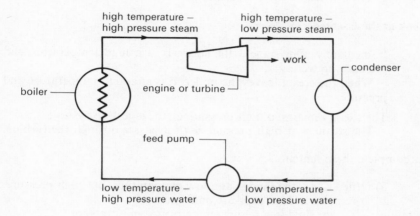

Steam plant

The water is pumped to the boiler by the feed pump. (Stage 1)
The water is heated in the boiler and becomes steam. (Stage 2)
The steam leaves the boiler. (Stage 3)
It passes to the engine or turbine. (Stage 4)
$\left\{ \begin{array}{l} \text{It passes through the turbine.} \\ \text{It gives up some of its energy.} \end{array} \right\}$ (Stage 5)
It leaves the turbine. (Stage 6)
$\left\{ \begin{array}{l} \text{It passes through the condenser.} \\ \text{It is condensed into water.} \end{array} \right\}$ (Stage 7)
The water leaves the condenser and passes to the feed pump. (Stage 8)

Now read this:

Stage 2 follows stage 1, so we say:

After the water is pumped to the boiler, it is heated and becomes steam.

The two parts of stage 5 are *simultaneous*, so we say:

As the steam passes through the turbine, it gives up some of its energy.

Stage 5 *precedes* stage 6, so we say:

Before the steam leaves the turbine, it gives up some of its energy.

Now complete these sentences:

1. Look and read:

    a) After the water is pumped to the boiler by the feed pump, ...

    b) After the steam leaves the boiler, ...

    c) Before the steam leaves the turbine, ...

    d) As ......, the steam gives up some of its energy.

    e) After ......, the steam passes to the condenser.

    f) As ......, the steam ....

    g) After it leaves the condenser, ...

2. Look at these sentences:

Immediately after stage 3 the steam is at a high temperature and pressure, so we say:

    When the steam leaves the boiler, it is at a high temperature and pressure.

The steam remains at high pressure up to stage 4, so we say:

    The steam is at high pressure *until* it passes through the turbine.

Now complete these sentences:

    a) When ......, the fluid is at a low temperature, but high pressure.

    b) It remains at high pressure until ....

    c) ......, the fluid is at a high temperature, and pressure.

    d) It remains at high temperature until ....

    e) When it leaves the turbine, ...

    f) ...... until it passes through the feed pump.

    g) ......, the fluid is at a low temperature and pressure.

3. Look at these sentences:

After the steam leaves the boiler, it passes to the turbine.
We can also say:
    After leaving the boiler, the steam passes to the turbine.

After the water is pumped to the boiler by the feed pump, it is heated and becomes steam.
We can also say:
    After being pumped to the boiler by the feed pump, the water is heated and becomes steam.

Before the steam leaves the turbine, it gives up some of its energy.
We can also say:
    Before leaving the turbine, the steam gives up some of its energy.

When the steam leaves the boiler, it is at a high temperature and pressure.

We can also say:

On leaving the boiler, the steam is at a high temperature and pressure.

As the steam passes through the turbine, it gives up some of its energy.

We can also say:

While passing through the turbine, the steam gives up some of its energy.

Now describe the sequence of actions in the following diagram in the same way:

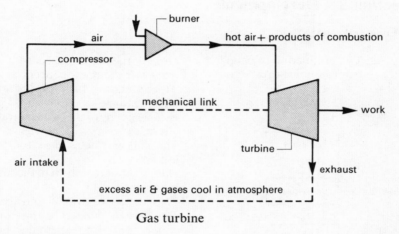

Gas turbine

4. Look at this sentence:

In the diagram of the steam plant, the water changes into a gas *during* evaporation (boiling).

What happens *during* condensation?

5. Look at these diagrams:

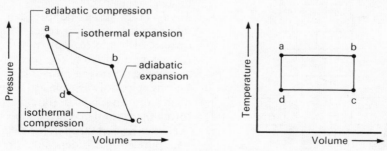

The Carnot cycle

Look at the relationship between the two diagrams and make true sentences from this table:

During	isothermal expansion, adiabatic expansion, isothermal compression, adiabatic compression,	the pressure	increases, decreases,

the volume	increases decreases	and the temperature	increases. remains constant. decreases.

Section 2 Development

6. Look and read:

The production of copper from sulphide ores.

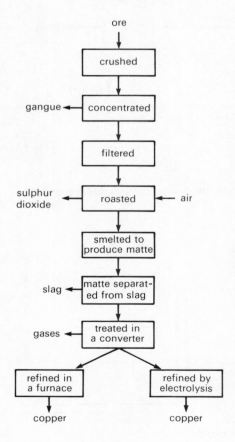

Initially the ore is crushed.
After this the ore is concentrated.
At this stage much of the earthy waste (gangue) is removed.
Following this the concentrate is filtered.
The concentrate is then roasted in hot air.
During this stage, much of the sulphur burns away.
Then the concentrate is smelted.
During this process a mixture of copper and iron sulphides (called matte) is formed.
At the same time a slag forms.
Subsequently, the matte is separated from the slag.
After this the matte is treated in a converter.
At this stage almost pure crude copper is formed. This is called blister copper.
Finally the blister copper can be refined either by being reheated in a furnace or by electrolysis. When it is refined in a furnace, the impurities are oxidised, leaving pure copper.
During electrolysis, blister copper forms the anode and pure copper forms the cathode. When an electric current is passed through the electrolyte, pure copper is deposited on the cathode.

Now say whether these statements are true or false. Correct the false statements:

    a) Much of the earthy waste is removed before the ore is concentrated.
    b) Much of the sulphur burns away while the concentrate is smelted.
    c) Most of the sulphur burns away after the matte is separated from the slag.
    d) The formation of matte and slag takes place during the fifth stage in the diagram.
    e) The matte is separated from the slag while it is treated in a converter.
    f) Copper is refined before being treated in a converter.
    g) Blister copper is formed in the converter and is subsequently refined to pure copper.
    h) The initial process is to concentrate the copper ore.
    i) The final process in the production of refined copper takes place in a converter.

7. Look at this diagram:

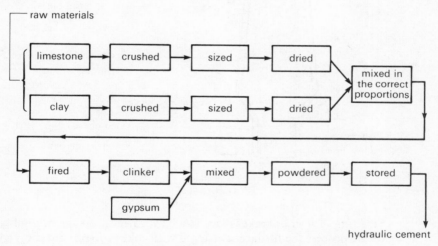

The stages in making cement

Complete these sentences with the following words and then put the sentences in the correct order as indicated by the diagram:

          then   after
          on   initially
          finally subsequently
          until

    a) ____ being fired, it emerges as clinker.
    b) ____ they are mixed together in the right proportions.
    c) This mixture is powdered and ____ stored ____ it is needed.

d) _____ cooling, the clinker is mixed with gypsum.
e) _____ the limestone and clay are crushed, sized and dried.
f) The mixture _____ passes to the kiln where it is fired at a maximum temperature of 3,000°C.

Section 3   Reading

8. Look at the diagram and read the passage:

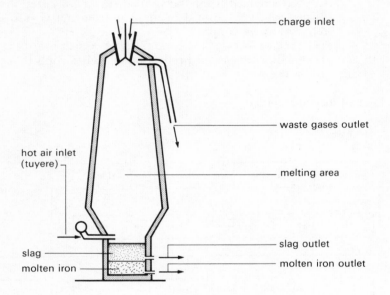

Blast furnace

Smelting of iron ore takes place in a blast furnace. At the beginning of the process the furnace is charged with coke (C), iron ore (Fe_2O_3) and limestone ($CaCO_3$), which are placed in alternate layers. Then a blast of hot air at high pressure enters through the tuyere at the bottom of the furnace. At this stage the coke burns to form carbon dioxide. As the hot carbon dioxide rises in the furnace, it meets more hot coke and is reduced to carbon monoxide. The carbon monoxide then reacts with the hot iron ore, forming iron and carbon dioxide. At the same time, the limestone absorbs earthy impurities from the ore, forming a liquid known as slag.

The molten iron and liquid slag drain to the bottom of the furnace, where they form two layers – the slag floats on top of the denser molten iron. The molten crude iron and slag pass out of the furnace at different levels. The main product of a blast furnace is known as pig or crude iron.

Now complete this chart:

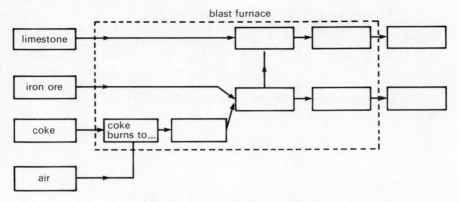

Reactions in the blast furnace

9. Answer these questions:

    a) What is the first stage of the smelting of iron ore?
    b) What is the second stage?
    c) Where is the tuyere located?
    d) When does the coke burn to form carbon dioxide?
    e) What happens to the carbon dioxide subsequently?
    f) Which reactions are simultaneous?
    g) Why does the slag float on top of the molten iron?
    h) What is the final stage of the smelting of iron ore?

Section 4   Listening

10. Listen to the description of the working of a refrigerator. As you listen label the diagram:

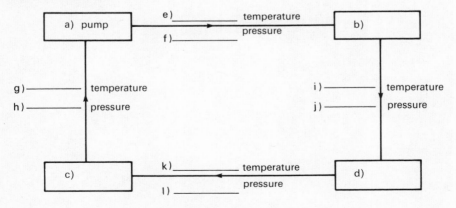

Refrigeration cycle

Now complete this description:

The working liquid in the refrigerator is ____. ____, it enters the
evaporator where it evaporates to form ____. At the same time the
temperature ____ and the low pressure vapour ____ ____ ____ ____ .
The pump ____ the pressure. ____, the vapour travels to the condenser
where the temperature ____, and the vapour ____ to form ____ . ____
it leaves the condenser, it passes to the ____ . ____ entering the
evaporator again, the pressure of the liquid freon ____ .

Unit B Revision

1. Read this:

The dry form of the Leclanché cell, which has an e.m.f. of 1·5V, is the primary cell most commonly used today.

The positive pole consists of a carbon rod surrounded by a mixture of carbon powder and manganese dioxide. This mixture acts as the depolariser, i.e. a mixture of substances that prevent polarisation and the consequential reduction in potential difference (p.d.). At the top of the carbon rod is a brass cap, which acts as the positive terminal. The carbon rod and the mixture of carbon powder and manganese dioxide which are contained in a porous gauze bag, are placed in a zinc container, and the space between the bag and the container is filled with the electrolyte, which consists of a paste of ammonium chloride and zinc chloride. The zinc container acts as the negative pole. The top of the cell is closed with a sealing compound.

Now match the letters with the names of the parts.

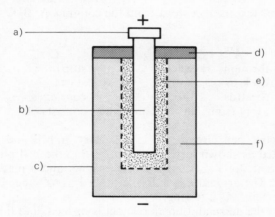

'Dry' primary cell

Complete the following sentences to match the idea in brackets:

a) The Leclanché cell ...... (measurement)
b) The positive pole ...... (structure)
c) The mixture of carbon powder and manganese dioxide ...... (location)
d) The mixture of carbon powder and manganese dioxide ...... (function)
e) The brass cap ...... (location)
f) The brass cap ...... (function)
g) The depolariser and the carbon rod ...... (location)
h) The zinc container ...... (function)

2. Look at this diagram:

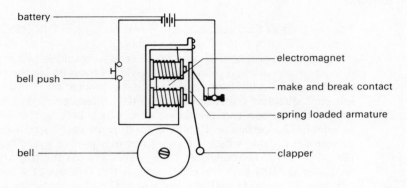

battery

electromagnet

bell push

make and break contact

spring loaded armature

bell

clapper

Electric bell and circuit

Now complete this text using a word from the columns A, B, C when indicated:

A	B	C
through	when	stops
back to	until	starts
towards	then	continues
away from	at the same time	consists of

An electric bell circuit __C__ a battery, a bell and a bell push connected as shown in the diagram. __B__ the bell push is closed, the current flows __A__ the electromagnet and a magnetic field is created. The armature is __B__ attracted __A__ the electromagnet and the bell __C__ to ring.

__B__ the make and break contact is opened, and this __C__ the flow of current __A__ the electromagnet. The magnetic force __C__ acting and the spring pulls the armature __A__ its initial position. __B__ the make and break contact closes, the current flows and the cycle __C__ again. The current __C__ to flow __B__ the bell push is released.

3. Read this:

Units of electricity

The coulomb is the unit of electrical charge. One coulomb consists of more than six million, million, million electrons. More precisely:

$$1 \text{ coulomb (Q)} = 6.24 \times 10^{18} \text{ electrons}$$

A material containing free electrons capable of moving from one atom to the next is a conductor. An electric current exists in a conductor when a charge is transferred from one point to another in that conductor. If a charge is transferred at the rate of one coulomb per second, a current of one ampere is flowing:

$$\text{Current in amperes (I)} = \frac{\text{Charge in coulombs (Q)}}{\text{Time in seconds (t)}}$$

An ammeter is an instrument for measuring the rate of flow of the current flowing in the circuit. An ammeter is always connected in series with the circuit.

The volt is the unit of electrical pressure or force that is needed to drive the current of electrons round the circuit. It is the force which uses unit energy to make unit charge flow.

$$1 \text{ volt} = 1 \text{ joule/coulomb}$$

The electromotive force (e.m.f.) is the difference in electrical pressure or potential measured across the terminals of a source of electrical energy such as a generator or battery on open circuit.

The potential difference is the difference in electrical pressure measured across two points in a circuit when a current is flowing through it. The potential measured across the terminals of a generator or battery forming part of a closed circuit is called the terminal potential difference and is always lower in value than the e.m.f. The terminal potential difference equals the sum of the potential differences across the various elements of the circuits, as in the diagram.

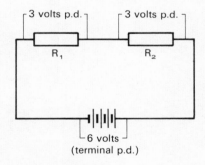

Series circuit

Voltage, both as e.m.f. and p.d., is measured by a voltmeter, which, unlike the ammeter, is connected in parallel with the circuit.

All conductors offer a resistance to the flow of electrons. Resistance is measured in ohms and can be calculated using the following formula derived from Ohm's Law:

$$R = \frac{V}{I}$$ where R is the resistance in ohms

         E is the p.d. across the resistor in volts
         I is the current flowing through the resistor in amperes

Say whether these statements are true or false. Correct the false statements.

a) Current equals the charge divided by the time in seconds.
b) The function of an ammeter is to measure the charge in a circuit.
c) An ammeter is connected in series with the other components in the circuit.
d) A volt equals the energy in joules multiplied by the charge in coulombs.
e) E.m.f. can be measured across the terminals of a battery under closed circuit conditions.
f) The function of a voltmeter is to measure both p.d. and e.m.f.
g) The voltmeter is connected in parallel with the other components in the circuit.
h) Resistance is measured in coulombs.

Unit 7   Measurement 2 Quantity

Section 1   Presentation

1. Look and read:

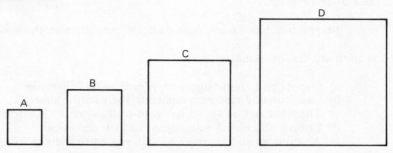

B is *slightly* bigger than A.
C is *considerably* bigger than A.
D is *much* bigger than A.
B is *approximately* $1\frac{1}{4}$ *times as* big *as* A.
C is *approximately three times as* big *as* A.
D is *approximately twelve times as* big *as* A.

Now look at this table:

	Density kg/m³	Melting point 0°C	Typical tensile strength N/mm²
Aluminium	2,700	660	90
Magnesium	1,700	651	248
Copper	8,950	1,083	340
Cast iron	7,860	1,535	116
Lead	11,300	327	15
Nickel	8,900	1,458	172
Titanium	4,500	1,725	216
Zinc	7,100	420	110

Identify these metals from the table:

a) This metal has a slightly higher density than nickel.
b) This metal has a slightly lower melting point than cast iron.
c) This metal has a tensile strength much greater than lead, but slightly lower than cast iron.
d) This metal is considerably lighter than aluminium but is much stronger in tension.
e) This metal has a considerably higher density than the other metals, but a much lower tensile strength.
f) This metal has a tensile strength approximately twice that of aluminium.

g) The melting point of this metal is approximately 1½ times as high as that of copper.
h) The density of this metal is approximately half that of copper.
i) This metal is three times as strong as zinc in tension.
j) These two metals have very nearly the same tensile strength.

2. Look at this example:

Magnesium has a *much higher* tensile strength *than* aluminium.

Now compare the following:

a) Copper and aluminium with regard to their densities.
b) Cast iron and zinc with regard to their tensile strength.
c) Titanium and nickel with regard to their tensile strength.
d) Copper and nickel with regard to their densities.
e) Aluminium and magnesium with regard to their melting points.

3. Look at this sentence:

Copper and nickel have very similar densities, but significantly different melting points and tensile strength.

Note: density and melting point can become plural, i.e. densities and melting points, but tensile strength normally remains singular.

Now compare the following in the same way:

a) Cast iron and zinc.
b) Aluminium and magnesium.
c) Nickel and cast iron.
d) Lead and zinc.

4. Look and read:

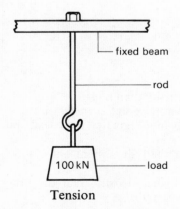

Tension

The rod is subjected to a tensile load. *Thus* it is said to be in tension. The material must be strong *enough* in tension to withstand the load. *In other words*, it must be *sufficiently* strong in tension to withstand the load. *i.e.* it must have *sufficient* tensile strength to withstand the load.

Now look at these diagrams and complete the sentences:

a)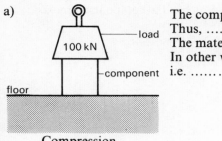

The component is subjected to a compressive force.
Thus, ...
The material ....
In other words, ...
i.e. ....

Compression

b)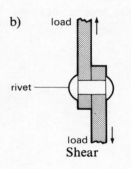

The rivet is subjected to a shear force.
Therefore ....
In other words, ...
i.e. ....

Shear

c)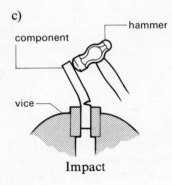

The component ...... impact load.
The material ...... tough to withstand the load.
In other words, ...
i.e. ....

Impact

5. Look again at the diagram on page 58 (Tension) and read these sentences:

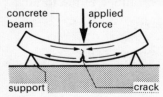

The beam has cracked because it is
not strong enough in tension.
The beam has cracked because it has
insufficient tensile strength.
The beam has cracked because of
lack of tensile strength.
The beam has cracked because the
load is *excessive*.

59

Now answer these questions in the same four ways:

a)

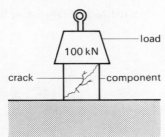

Why has the component cracked?

b)

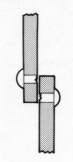

Why has the rivet broken into two parts?

c)

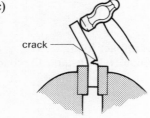

Why has the component cracked?

6. Read this:

If materials are subjected to excessive load, they may break. Therefore when a material is used in a building e.g. in a column, there is a maximum load which can be safely supported for any given cross-section. If the force is known, the minimum size of cross-section which will support it can be found by substituting in the formula

$$A = \frac{P}{\sigma}$$

where A = area
P = force
and σ is a constant called
basic allowable stress

Now look at this table:

The approximate basic allowable stress (in compression) for eight common materials

Material	Basic allowable stress (N/mm²)
Grey cast iron	120
Mild steel	155
Alloy steel	390
Aluminium	30
Pine	10
Brick	1·3
Concrete	5·3
Copper	60

For each of the materials in the table, calculate the minimum cross-sectional area which will support a compressive load of 20,000 N (20 kN).

Example: Grey cast iron: $A = \dfrac{P}{\sigma} = \dfrac{20,000}{120} = 166 \cdot 67 \, mm^2$

Say whether the following are true or false. Correct the false statements.

    a) The minimum cross-sectional area of grey cast iron which will support 20,000 N is $166 \cdot 67 \, mm^2$.

    b) The minimum cross-sectional area of mild steel which will support 20,000 N is $150 \, mm^2$.

    c) The minimum cross-sectional area of the alloy steel which will support 20 kN is $512 \cdot 82 \, mm^2$.

    d) The minimum cross-sectional area of aluminium which will support 20 kN is $666 \cdot 67 \, mm^2$.

    e) A pine beam with a cross-sectional area of $500 \, mm^2$ will support a comprehensive load of 500 N.

    f) A brick with a cross-sectional area of $15 \cdot 384 \, mm^2$ will support a load of 200 N.

    g) A concrete beam with a cross-sectional area of $188 \cdot 679 \, mm^2$ will support a load of 1 kN.

    h) A copper rod with a cross-sectional area of $666 \cdot 67 \, mm^2$ will support a load 50 kN.

7. Read this:

An engineer is designing a series of columns which must support a compressive load of up to 30 kN. The table below shows the different materials he could use. The second column in the table shows the minimum size of cross-section that will support the load safely. The third column in the table shows the size of the available members for each material. The engineer must consider the following data when making a decision:

Minimum cross-sectional area: 100 mm²
Maximum cross-sectional area: 500 mm²
Maximum cost per metric tonne: £200
Minimum degree of corrosion resistance: moderate
Maximum number of man hours to erect: 8

Now look at the table:

Material	Minimum safe cross-section mm²	Available cross-section mm²	Cost per tonne £	Corrosion resistance	Man hours needed
Alloy steel	77	77	175	low	7
Aluminium	1,000	1,000	300	high	4
Brick	230·77	231	25	high	9
Copper	500	500	520	high	5
Duralumin	333	300	250	high	4
Grey cast iron	250	275	80	low	5
Mild steel	194	194	100	moderate	5
Pine	3,000	2,500	35	low	7

Now say which material the engineer will choose, and which he will not:

Example: Advantages of pine: Pine is *cheap enough*, and can be erected *quickly enough*.

Disadvantages: *On the other hand*, it has *insufficient* corrosion resistance, and the available cross-sectional area is *too great* and unsafe.

Conclusion: *Therefore* the engineer will not choose pine.

Section 3 Reading

8. Read this:

The demand for electricity varies considerably during twenty-four hours. The base load is the amount of electricity which is always needed and is therefore constant day and night. It is supplied by heavy duty generators which are powered by coal, oil or nuclear fuel. The quantity of electricity which they supply is inflexible, i.e. it cannot be varied easily. This is because the generators take a considerable period of time to reach maximum capacity when they are started from rest. Plant of this kind is most economical if it is run at a constant load and full capacity.

Fluctuating load is the load which occurs mostly during daylight hours. The load is met by medium sized conventional generators (i.e. steam driven) or by hydro-electric generators. These can produce electricity quickly, especially the hydro-electric type. At the moment, hydro-electric power plants provide the most economical and flexible supply as the energy source is cheap and electricity can be produced in a few minutes. The generators which provide this supply are run below full capacity to allow for emergencies, such as unexpected cold weather. During the peak period from 1500 to 1800 hours, the demand for electricity exceeds the amount provided by the base and fluctuating supplies. The extra load is supplied by hydro-electric and diesel generators. Like hydro-electric generators, diesel generators have the capacity to produce electricity very quickly. The extra demand on a cold day is met by running the generators which supply the fluctuating demand at greater capacity and by using the generators which supply the evening peak power and by special emergency plant.

9. Look at this graph. Match the letters with these labels:

the base load, the fluctuating load, the peak load

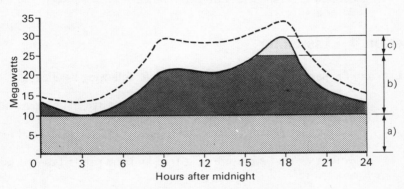

Electricity supply load/time graph

10. Now answer these questions:

    a) In the graph what does the dotted line show?
    b) Why are heavy duty generators used only for the base load?
    c) What advantage do hydro-electric generators have?
    d) During which period are both hydro-electric and diesel generators used?
    e) What is the similarity between hydro-electric and diesel generators?
    f) When is every kind of generator used?

11. Below is a summary of the reading passage, but the points are listed in the wrong order. Put them in the right order:

    a) The advantages of steam-driven or hydro-electric generators.
    b) What happens during a cold day.
    c) The definition of base load.
    d) The definition of fluctuating load.
    e) What happens during the peak period.

f) How the base load is supplied.
g) How the fluctuating load is supplied.
h) The advantage of diesel generators.

12. Complete this passage, using these words:

extra   to   meet   peak   insufficiently   sufficient   enough

Heavy duty generators have ____ capacity to supply the base load, but they are ____ flexible to be used economically for supplying any additional load. Steam-driven generators or hydro-electric generators can increase their output of electricity quickly and cheaply ____ to ____ most fluctuating demands. These generators are operated at low power during the day, but at ____ periods their output is increased, and additional diesel and hydro-electric generators can be brought into operation. By this means ____ electricity can be generated ____ meet the demand which occurs between 1500 and 1800 hours.

Section 4   Listening

13. Listen to the passage, and make notes under the following headings:

The point at which the temperature of a tungsten filament remains constant.
The difference between a tungsten filament, and the wires used in an electric circuit.
The function and structure of a fuse and the properties necessary for fuse wire.

Now answer these questions:

a) At what point does the temperature of a tungsten filament in which electric current is flowing remain constant?
b) Why must the wires used to connect an electric circuit be relatively thick?
c) What is the function of a fuse, and what does it consist of?
d) What are the properties of the wire in a fuse which enable it to break a circuit?
e) In a lighting circuit why must the fuse wire heat up at a slower rate than the filaments in the lamps?

Unit 8　Process 3 Cause and Effect

Section 1　Presentation

1. Look at this diagram:

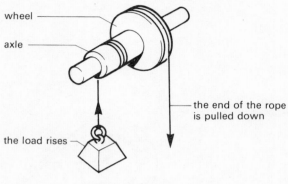

wheel

axle

the end of the rope
is pulled down

the load rises

Wheel and axle

Look at these examples:

The end of the rope is pulled down $\left\{ \begin{array}{l} \textit{with the result that} \\ \textit{and, as a result,} \end{array} \right\}$
the load rises.
Pulling the rope down *causes* the load *to* rise.
The rope is pulled down *thereby causing* the load to rise.

Now complete these sentences:

　a) The rope moves and, as a result, the wheel ＿＿.
　b) The movement of the rope causes the wheel ＿＿.
　c) The rope moves, thereby causing ....

2. Look at this diagram:

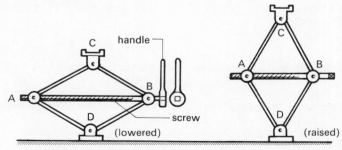

C

handle

B

A

D

screw

(lowered)

C

A　　B

D

(raised)

Screw jack: a device for raising heavy weights

Complete these sentences:

      a) The handle is turned and, as a result, the screw _____ .
      b) The rotation of the screw causes A ....
      c) A moves towards B with the result that ....
      d) C rises in respect to D, thereby causing ....

Now write four sentences like those above to describe how the screw jack
is lowered.

3. Look and read:

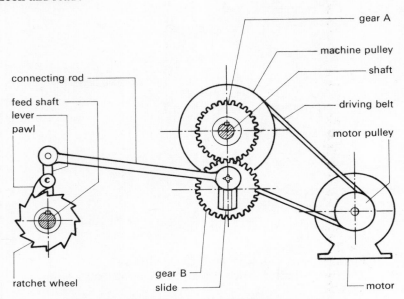

Pawl and ratchet intermittent feed mechanism

In this mechanism the continuous rotation of the shaft is converted
into intermittent rotation of the feed shaft. It is often used in shaping
machines, where the intermittent movement of the feed shaft causes
the table to move forward in steps.

Now write a description of how the feed mechanism operates by joining the
correct half-sentences:

a) The drive from the motor causes	the connecting rod move backwards and forwards.
b) The shaft rotates	thereby causing the slide to rotate.
c) As a result of the rotation of gear A	rotational movement to be transmitted into linear motion.
d) Gear B rotates	and as a result the table moves.
e) Rotation of the slide makes	thus causing gear A to rotate in a clockwise direction.

f) The connecting rod moves backwards and forwards	thus making the ratchet wheel rotate intermittently.
g) The pawl moves backwards and forwards	gear B rotates.
h) The intermittent rotation of the ratchet wheel causes	thus pushing the pawl backwards and forwards.
i) The feed shaft rotates	the shaft to rotate in a clockwise direction.
j) The feed mechanism enables	the feed shaft to rotate.

4. Look at these diagrams:

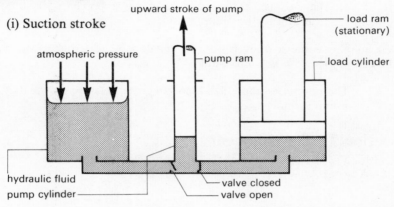

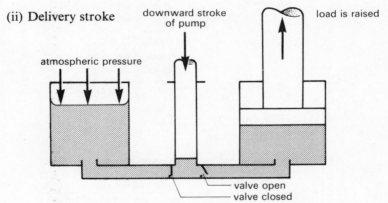

Principle of hydraulic jack

Look at diagram (i) and read this:

> The left-hand valve is open. This *allows* the fluid in the cylinder on the left *to enter* the pump cylinder. The right-hand valve is closed. This *prevents* the fluid in the load cylinder *from returning* to the pump cylinder.

Now describe the valves and their functions in diagram (ii).

67

Now re-order these sentences to describe the sequence of events during the upward stroke of the pump (the suction stroke):

a) The opening of the left-hand valve allows the fluid to enter the pump cylinder.
b) The pump rises, thus creating a partial vacuum in the pump cylinder.
c) The back pressure from the fluid in the load cylinder causes the right-hand valve to close.
d) The atmospheric pressure on the fluid in the cylinder on the left causes the left-hand valve to open.
e) The closing of the right-hand valve prevents the fluid from returning to the pump cylinder.

Describe the sequence of events during the downward stroke of the pump (the delivery stroke). Begin like this:

The pump descends thus increasing the pressure on the fluid in the pump cylinder ....

Section 2 Development

5. Look at this table:

Metal	Chemical symbol	The principal effects when added to steel
Manganese (15%)	Mn	the creation of self-hardening properties when the steel is subjected to abrasion or rubbing
Nickel	Ni	an increase in strength toughness and ductility
Chromium	Cr	an increase in hardness and an improvement in corrosion resistance
Tungsten	W	an increase in retention of hardness at high temperatures
Vanadium	V	an increase in toughness and a reduction in the risk of fatigue failure

The principal effects of alloying elements on alloy steels

Look at these sentences:

The presence of nickel $\begin{Bmatrix} results\ in \\ brings\ about \end{Bmatrix}$ an increase in strength, ductility and toughness.

We can also say:

An increase in strength, ductility and toughness $\begin{Bmatrix} is\ brought\ about\ by \\ results\ from \end{Bmatrix}$ the presence of nickel.

Make similar sentences about the other metals.

Which alloying element would be appropriate to add to a steel for each of the following applications? Give reasons, referring to the table.

Example: Application: Cutlery (knives, forks etc.)

Answer: Chromium, because its effect on steel is to improve resistance to corrosion.

a) The blade of a bulldozer.
b) A cutting tool material suitable for operation without a coolant.
c) A heavily loaded back axle of a lorry.
d) A valve spring subjected to long periods of rapid change of stress.
e) A tank used for storing mildly corrosive chemicals.

6. Read this:

The effect of adding carbon to plain carbon steel is to reduce the ductility considerably. This continues until the carbon content reaches 0.87%. Subsequently the ductility falls more slowly.

Another effect is to increase the hardness of the steel. The hardness of the steel increases as the carbon content increases.

The presence of carbon causes the tensile strength to increase rapidly, up to 0.87%. After the carbon content reaches 0.87%, the strength slowly falls off.

Now copy and complete these graphs:

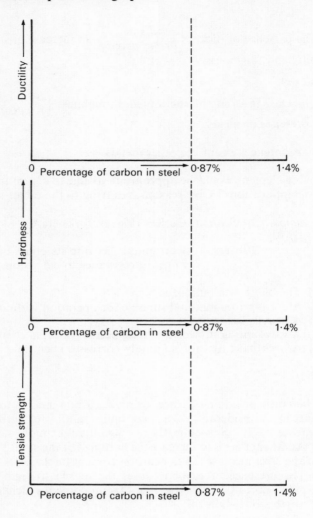

The effect of carbon content on the properties of plain carbon steels

Look at this sentence:

If the carbon content is increased up to 0·87%, the ductility of the steel will decrease rapidly.

Now make similar statements with information from the graphs.

Section 3   Reading

7. Look at this diagram:

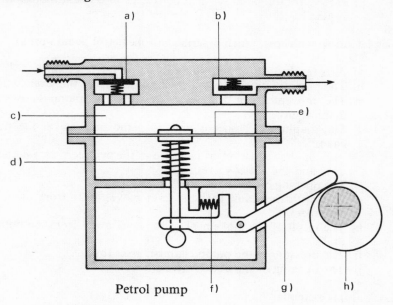

Petrol pump

Now read this:

When the car engine is running, the cam rotates against the lever. Because of the rise and fall of the periphery of the cam, the pressure on the lever varies, alternately pushing it towards the pump and allowing it to return. When the outer end of the lever is pushed towards the pump, the diaphragm is pulled down, thus reducing the pressure in the pump chamber. As a result of this, the inlet valve opens and the outlet valve shuts. Petrol is then sucked from the tank into the pump.

As the cam turns further, the pressure on the outer end of the lever is reduced. Consequently the lever is forced away from the pump by the lever spring. As the lever no longer holds the diaphragm down, the diaphragm spring is able to return the diaphragm to its original position. The pressure in the chamber is thus increased, causing the inlet valve to shut and the outlet valve to open. Petrol is then forced out of the pump to the carburettor.

Now match the letters with the names of the parts.

8. Answer these questions:

    a) What is the function of the cam?
    b) What kind of movement does the lever make?
    c) What happens when the diaphragm is pulled down?
    d) What are the functions of the inlet and outlet valves?

e) What is the function of the lever spring?

f) What is the function of the diaphragm spring?

g) How is rotational movement converted into linear motion in the system?

9. Complete these sentences, which describe how the petrol pump works:

a) As a result of the ......, the pressure on the lever varies.

b) The pressure on the lever varies, thus causing the lever ....

c) The movement of the lever towards the pump causes the diaphragm ....

d) The diaphragm descends and ____ the pressure in the pump chamber ....

e) ...... of the reduction of pressure in the cylinder, the inlet valve ____ and the outlet valve ____.

f) The opening of the inlet valve allows ....

g) ...... with the result that the lever moves away from the pump chamber.

h) The result of the movement of the lever away from the pump is that ....

i) The pressure in the pump chamber rises thus ....

j) The opening of the outlet valve allows ....

10. Look at this example:

The rise and fall of the periphery *causes* the pressure on the lever to vary.

The variation in pressure on the lever $\left\{ \begin{array}{l} \textit{is due to} \\ \textit{is caused by} \end{array} \right\}$ the rise and fall of the periphery of the cam.

Now make a similar change to these sentences:

a) The variation in pressure on the lever causes it to move intermittently.

The intermittent movement of the lever ....

b) The rise of the diaphragm causes the pressure in the pump chamber to increase.

The increase in pressure in the pump chamber ....

c) The reduction in pressure on the lever causes it to move away from the pump chamber.

The movement ......

d) In the feed mechanism on page 66 the action of the motor causes the shaft to rotate.

The rotation ......

e) The rotation of the gears causes the connecting rod to move backwards and forwards.

The backwards and forwards movement ......

f) The backwards and forwards movement of the connecting rod causes the pawl to oscillate.

The oscillation ......

g) The oscillation of the pawl causes the ratchet wheel and the feed shaft to rotate intermittently.
   The intermittent ......

h) The intermittent rotation of the feed shaft causes the table to move in steps.
   The stepped movement ......

Section 4   Listening

11. The diagrams below are in the wrong order. Listen to the text and list them in the correct order.

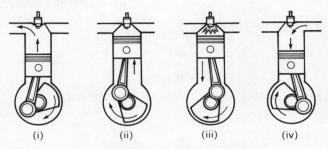

Four stroke Otto cycle

12. Now complete these sentences:

a) In stage (i) the piston descends and as a result ....

b) As a result of ......, the air/petrol mixture enters the cylinder.

c) In stage (ii) the crankshaft rotates and, consequently, the piston .... The inlet valve ....

d) As a result of the rise of the piston, ...

e) In stage (iii) ...... with the result that a large volume of gas is produced.

f) The increased pressure in the cylinder causes ....

g) In stage (iv) ....

Unit 9   Measurement 3 Ratio and Proportion

Section 1   Presentation

1. Look at these diagrams:

Gear ratios

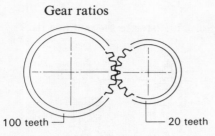

100 teeth — — 20 teeth

Example:
The ratio of the gear teeth is
    100:20 or 5:1
In other words the small gear (20t)
rotates 5 times as fast as the large
gear (100t)

Now answer these questions and complete the sentences:

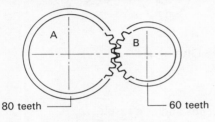

80 teeth — — 60 teeth

a) What is the ratio of gear A to
gear B?
In other words gear B ....

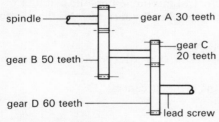

spindle — gear A 30 teeth

gear C 20 teeth

gear B 50 teeth

gear D 60 teeth — lead screw

b) What is the ratio of A to B?
What is the ratio of C to D?
What is the overall ratio (i.e. A
to D)?
In other words the lead
screw ....

2. Look and read:

The function of a lifting machine is to magnify an effort in order to
raise a load considerably larger than one that could be raised by the
unaided effort. No machine is 100% efficient, and therefore some
of the effort must be used in driving the machine.

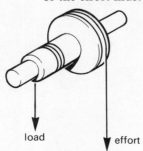

load   effort

The Mechanical Advantage (M.A.)
or force ratio can be calculated from
the formula:

$$M.A. = \frac{Load}{Effort}$$

The Velocity Ratio (V.R.) can be calculated from the formula:

$$\text{V.R.} = \frac{\text{distance moved by effort in unit time}}{\text{distance moved by load in unit time}}$$

or

$$\text{V.R.} = \frac{\text{distance moved by effort}}{\text{distance moved by load}}$$ (This shortened formula is more correctly called a displacement ratio or a movement ratio.)

Efficiency (η) can be calculated from the formula:

$$\eta = \frac{\text{work output}}{\text{work input}} = \frac{\text{Load} \times \text{distance moved by load}}{\text{Effort} \times \text{distance moved by effort}}$$

or

$$\eta = \frac{\text{M.A.}}{\text{V.R.}} \text{ or, expressed as a percentage, } \frac{\text{M.A.} \times 100}{\text{V.R.}}$$

Look at this example:

The mechanical advantage equals the load $\left\{\begin{array}{l}over\\divided\ by\end{array}\right\}$ the effort.
This means that the mechanical advantage is the ratio between the load and the effort.

Now complete these:

a) The velocity ratio equals ....
b) This means that ....
c) Efficiency equals ....
d) This means ....
e) As well as being expressed as a ____, efficiency can also be expressed as a ____.

Now look at this diagram:

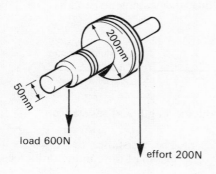

load 600N

effort 200N

$$\text{Mechanical Advantage} = \frac{\text{Load}}{\text{Effort}}$$

$$= \frac{600\text{N}}{200\text{N}}$$

$$= 3{:}1$$

Using the formula complete these:

  f) The mechanical advantage is ....
  g) The velocity ratio is ....
  h) The efficiency is ...... or ...... %.
  i) The load is ____ times as large as the effort.
  j) The load moves ____ times further than the effort.

Section 2   Development

3. Look and read:

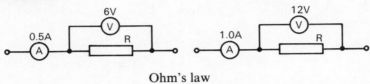

Ohm's law

Given that the circuit resistance remains constant, the current *increases* as the potential difference *increases*. Thus, *the greater* the potential difference, *the greater* the current.
In other words the current is *directly* *proportional* to the potential difference.

Relationship between conductor thickness and resistance

The resistance of a conductor *decreases* when the cross-sectional area *increases*. Thus, *the smaller* the cross-sectional area, *the greater* the resistance.
In other words, the resistance is *inversely proportional* to the cross-sectional area.

Look at the diagrams and say whether the statements are true or false. Correct the false statements.

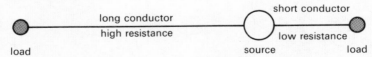

Relationship between conductor length and resistance

a) The longer the conductor, the lower the resistance.
b) The resistance is directly proportional to the length of the conductor.

c) The resistance decreases as the length of the conductor increases.

76

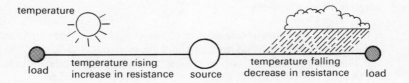

load — temperature rising / increase in resistance — source — temperature falling / decrease in resistance — load

Relationship between conductor temperature and resistance

d) With metal conductors, the higher the temperature, the higher the resistance.

e) The resistance of metal conductors is inversely proportional to the temperature.

f) As the temperature increases, the resistance of metal conductors increases.

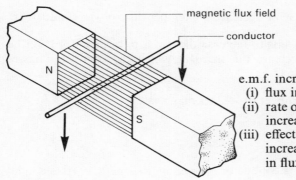

magnetic flux field

conductor

N

S

e.m.f. increases if:
  (i) flux increases
  (ii) rate of cutting flux by conductor increases
  (iii) effective length of conductor increases (length of conductor in flux field)

g) The induced e.m.f. is directly proportional to the magnetic flux.

h) The faster the conductor cuts the flux the greater the induced e.m.f.

i) The induced e.m.f. is directly proportional to the time taken by the conductor to cut the flux.

j) As the length of the conductor increases, the induced e.m.f. decreases.

4. Now complete these sentences:

a) The ____ the cross-sectional area of the wire, the ____ the resistance.
b) As the cross-sectional area of the wire, ...
c) The resistance is ____ proportional to the length of the conductor.
d) As the temperature decreases, ... of metal conductors ____ .
e) The ____ of metal conductors is ....
f) In a generator, the ____ the magnetic flux, ...
g) In a generator, the induced e.m.f. is ....
h) In a generator, the ____ the length of the conductor, ...

5. Answer these questions:

a) An engineer wants to reduce the resistance in a circuit.
Will he increase or decrease the length of the conductors?
Why?
b) Will he use thinner or thicker wires?
Why?
c) Will he increase or decrease the temperature of the conductors?
Why?
d) An engineer wants to increase the e.m.f. in a generator. Will he increase or decrease the speed at which the conductor passes through the flux?
Why?
e) Will he use stronger or weaker magnets?
Why?

6. Read this:

The resistance of a conductor *depends on* its length, cross-sectional area and its temperature.

Now look again at the last diagram and make true sentences from this table:

The strength of the induced e.m.f. The direction of the induced current	depends on	the strength of the magnetic flux. the direction of movement of the conductor. the rate of cutting of the flux by the conductor. the direction of the magnetic flux.

Section 3 Reading

7. Read this:

Concrete consists of a number of substances mixed in certain fixed proportions, depending on the result required. One of the commonest mixtures is called standard mass concrete, which is made up of cement, fine aggregate (sand) and coarse aggregate (small stones) in
5   the proportions 1:2:4. Thus for every part of cement, there are two parts of fine aggregate and four parts of coarse aggregate.

The strength of concrete depends on the quantity of water used in relation to the cement content. In other words, the lower the proportion of water to cement, the greater the strength. After the optimum
10   amount of water is reached, any excess of water, however slight, reduces the strength of the concrete. The relationship between the quantity of mixing water and the amount of cement in a concrete mix is known as the water-cement ratio and is expressed as $\dfrac{W}{C}$, either by volume or by mass or by the number of litres of water per 50 kg
15   bag of cement.

Thus, if expressed by volume, 1 m³ of water per 1 m³ of cement equals a ratio of $\dfrac{1}{1} = 1$.

If expressed by mass, this ratio becomes

$$\frac{1{,}000\text{ kg (mass of 1 m}^3\text{ of water)}}{1{,}442\text{ kg (mass of 1 m}^3\text{ of cement)}}$$

20   If 26·85 litres of water are used per 50 kg of cement the ratio is

$$\frac{26{\cdot}85\text{ kg}}{50\text{ kg}} = 0{\cdot}54$$

After solidifying, the strength of concrete depends on its age. Its strength can be further increased by curing while it is hardening. Curing is the keeping of concrete moist during the first week or
25   month of its age. The longer the concrete is kept moist, the stronger and the more durable it becomes.

Say whether these statements are true or false. Correct the false statements.

    a) In standard mass concrete there is four times as much coarse aggregate as fine aggregate.
    b) There is twice as much coarse aggregate as cement.
    c) The strength of concrete is inversely proportional to the water/cement ratio.
    d) As the water/cement ratio increases, the strength of concrete increases.
    e) A slight excess of water does not have an effect on the strength of concrete.
    f) The water/cement ratio is the mass of water divided by the mass of the cement.

g) The strength of concrete is inversely proportional to length of time it is cured.

h) The longer concrete is cured, the stronger it becomes.

8. Now answer these questions:

a) In line 11 what does 'the relationship' refer to?

b) In lines 22/3 what does 'its strength' refer to?

c) What is standard mass concrete?

d) Define the water/cement ratio.

e) If there is $2\,m^3$ of water and $3\,m^3$ of cement what is the ratio by volume and by mass?

9. Below is a list of points contained in the reading passage, but they are listed in the wrong order. Put them in the correct order.

a) The relationship of the amount of water and the amount of cement used to make concrete is known as the water/cement ratio.

b) After solidification, concrete can be strengthened by curing.

c) The ratio of the cement to fine aggregate to coarse aggregate in standard mass concrete is 1:2:4.

d) The water/cement ratio can be expressed by mass, volume or by the number of litres used per bag of cement.

e) The strength and durability of concrete are directly proportional to the length of time it is cured.

f) The strength of concrete is inversely proportional to the proportion of water to cement after the optimum amount of water is reached.

g) After solidification, the strength of concrete is affected by age.

10. In this passage the writer discusses the following:

(1) the composition of concrete

(2) the relationship between the water/cement ratio and the strength of concrete

(3) other factors affecting the strength of concrete

Under (1) the composition of concrete the writer deals with these points:
a) the composition expressed as a ratio
b) the composition expressed as the number of parts of different materials

Now list the points made under these headings:

(2) the relationship between the water/cement ratio and the strength of concrete

(3) other factors affecting the strength of concrete

Section 4   Listening

11. Copy this graph and carry out the instructions which you will hear.

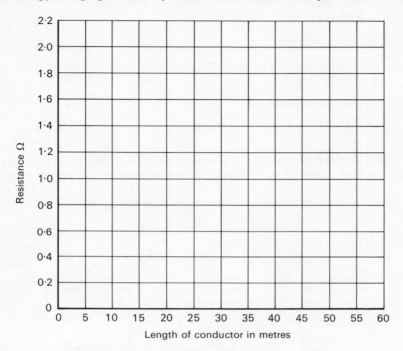

Graph of conductor length and resistance

Unit C   Revision

1. Look at this diagram:

Dimensions in millimetres

Shaft tolerance

40 ± 0.05 means that the component must not be bigger than 40·05 mm (the upper limit) and must not be smaller than 39·95 mm (the lower limit).

The tolerance is the difference between the upper limit and the lower limit, and in this case is 0·10 mm.

Now say what these mean:

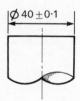 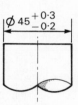

Look at the diagram of shaft tolerance again. Now say whether the following shafts are satisfactory for the component in the diagram:

a) A shaft of diameter 40·07 mm.
b)   ,,   ,,   39·90 mm.
c)   ,,   ,,   41·05 mm.
d)   ,,   ,,   39·95 mm.
e)   ,,   ,,   38·90 mm.
f)   ,,   ,,   39·96 mm.
g)   ,,   ,,   40·09 mm.
h)   ,,   ,,   40·04 mm.

Make true statements from this table to support your decision:

This shaft is	considerably slightly	larger smaller	than the	upper lower	limit
	within limits				

and is, therefore,	too small (undersize). too big (oversize). satisfactory.

2. Look and read:

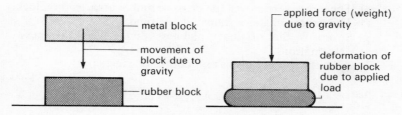

The effect of gravity

The diagram shows a metal block supported on a rubber block. The force of gravity acts on the mass of the metal block, thereby causing it to apply a downward force to the rubber. The downward force of the metal causes the rubber to change shape, and this change of shape is known as deformation. The change in thickness divided by the original thickness is a ratio and is referred to as strain.

$$\text{Strain} = \frac{\text{Change in thickness or length (x)}}{\text{Original thickness or length (l)}}$$

The rubber block resists the downward force of the metal by an internal reaction force, distributed over the cross-sectional area of the block. This internal reaction force is called stress. As long as the metal block rests on the rubber block, the rubber is said to be in a state of stress. The more the rubber is distorted by the load, the greater will be the stress. Provided the downward force is not excessive, the stress will eventually balance the load, and no further deformation will take place. Stress can be calculated by dividing the force by the area of the rubber block, i.e.

$$\text{Stress} (\sigma) = \frac{\text{Force}}{\text{Area of rubber block}}$$

With elastic materials, stress and strain can be related to give the modulus of elasticity (Young's modulus) 'E', which is the stress divided by strain.

$$\text{Modulus of elasticity (E)} = \frac{\text{Stress}}{\text{Strain}}$$

Now write definitions of the following:

    a) Deformation
    b) Strain
    c) Stress
    d) Young's modulus

3. Say whether these statements are true or false. Correct the false statements.

    a) The rubber changes shape as a result of the downward force of the metal block.
    b) The greater the decrease in thickness in the rubber block, the lesser the strain.
    c) Stress is an external force.
    d) Stress is directly proportional to the applied force.
    e) The stress varies over the cross-sectional area of the block.
    f) The rubber changes shape all the time.
    g) Strain is the original thickness divided by thickness after deformation.
    h) 'E' is the symbol given to modulus of plasticity.

4. Read this:

The heat treatment of steel
The properties of steel can be varied by heat treatment. This is because of the structural changes which take place during the heating and cooling of the steel.

If a plain carbon steel containing 0·6% carbon is heated, its internal structure begins to change at about 700°C and it is completely changed into a new form at about 800°C. The former temperature is called the lower critical point, and the latter temperature is called the upper critical point. These temperatures vary according to the carbon content of the steel.

Annealing
Annealing makes the steel previously described softer and more ductile. After annealing it can be cold-worked without the danger of fracture. The steel is heated to 30°–50°C above the upper critical point and held at this point for a period of time which depends on the thickness of the metal. It is then cooled very slowly, usually in the furnace.

If the steel is heated to a temperature above the correct annealing temperature, or if it is kept too long at the annealing temperature, this results in excessive grain growth, and the metal becoming weak and brittle. The effect of overheating can be remedied by allowing the metal to cool down and then annealing the metal again at the correct temperature.

Normalising
When the 0·6% carbon steel is normalised, it is heated to a point 30°–50°C above the upper critical point and then cooled slowly in

still air. This makes the steel harder and stronger than when it is annealed but less ductile. It also makes it more suitable for machining as annealed steel tends to tear and leave a poor finish.

Hardening
When the 0·6% carbon steel is hardened, it is heated to 30°–50°C above the upper critical point and cooled rapidly by quenching in water, brine or oil. This makes the metal hard but brittle.

Tempering
Hardened steel can be made less brittle by tempering. This process consists of heating the metal to a temperature between 210° and 290°C and then quenching it. This quenching is less rapid than with hardening. The result of tempering is to make the metal tougher and more ductile. On the other hand, the hardness and tensile strength are reduced.

Now copy and complete the following chart to show the effects of the different rates of cooling on metal. The first one is done for you:

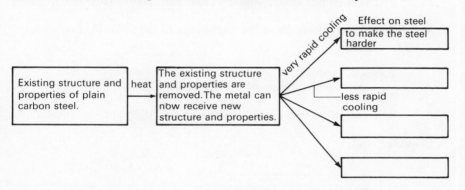

5. What heat treatment process would be used prior to the following operations? Give reasons:

    a) a metal part which needs to be bent
    b) machining a large forging on a lathe
    c) grinding the cutting edge on a newly made chisel (two processes)
    d) cold drawing a wire from 0·6% carbon steel

Unit 10   Measurement 4
Frequency, Tendency, Probability

Section 1   Presentation

1. Look at this example:

> *Most* cutting tools are made of steel. (usually)
> Cutting tools are *usually* made of steel.

Now change the following into statements of frequency using the word in brackets:

> a) A few cutting tools are made of ceramic. (occasionally)
> b) Some cutting tools are made of diamond. (sometimes)
> c) Most cutting tools are made of high speed steel. (usually)
> d) No cutting tools are made of cast iron. (never)
> e) Many cutting tools are made of high carbon steel. (often/frequently)

2. Make similar statements about the application of the following materials to gears:

> *Example:* metal
>   Gears are *usually* made of metal.

> a) steel   b) wood   c) plastic materials   d) brass   e) ceramic

3. Look and read:

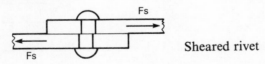

Sheared rivet

Observation:
The rivet has sheared because its diameter is too small to resist the shearing force Fs.
We can make the following *generalisation*:

Rivets $\left\{ \begin{array}{c} \text{\textit{tend to}} \\ \text{\textit{are likely to}} \end{array} \right\}$ shear if their diameter is too small to resist the applied load.

Now look at these diagrams:

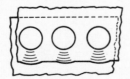

Crushed metal

The diameter of the rivet is too large compared with the thickness of the metal.
a) What observation can we make?
   The rivets have bulged and crushed the metal in front of them.
b) What generalisation can we make?
   Rivets tend to ....

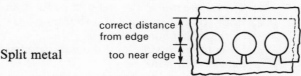

Split metal

The rivets are too near the edge of the plate.
c) What observation can we make?
d) What generalisation can we make?

Section 2   Development

4. Look at this graph:

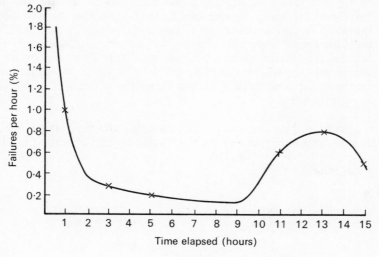

Failure rate

This graph shows the failure rate of components produced in a given factory.

Look at these statements:

The failure rate at the end of the first hour is approximately 1 %. This means that ten out of a thousand components *are liable to* fail at the end of the first hour of operation.
In other words, the *chances* of a component failing at the end of the first hour of operation are *one in a hundred*.

Now say whether the following statements are true or false. Correct the false statements.

     a) Four out of a thousand components are liable to fail at the end of the third hour of operation.
     b) The chances of a component failing at the end of the third hour are one in three hundred.
     c) The failure rate at the end of the fifth hour is approximately 2%.
     d) Four components out of a thousand are liable to fail at the end of the fifth hour.
     e) Six components out of a thousand are liable to fail at the end of the eleventh hour.
     f) The failure rate at the end of the thirteenth hour is approximately 0·8%.
     g) The chances of a component failing at the end of the fifteenth hour are one in a hundred.
     h) Of ten thousand components produced, five hundred would fail by the end of the fifteenth hour.

5. Read this:

In a given system the chances of a component failing after the first five thousand hours are one in a thousand. During the next three thousand hours the chances of failure uniformly decrease until after eight thousand hours the chances are one in two thousand. The failure rate remains constant for the next four thousand hours. It then enters the wearout period and the chances of the component failing in the next three thousand hours increase uniformly to one in eight hundred.

Now draw a graph to plot the changes in failure rate as a percentage. Make the vertical axis the failure rate and the horizontal axis the operating hours.

6. Look at this graph:

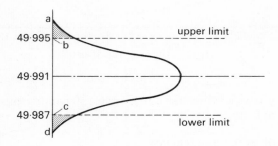

Component size distribution curve

A given machine has to produce components with the diameter within the limits 49·987 to 49·995 mm. The graph above shows the distribution of the size of the components.

Look at these examples:

> The *probability* of the machine producing a component the diameter of which is within the range 49·991 to 49·992 mm is *very high*.
> The *probability* of the machine producing a component the diameter of which is within the range 49·992 to 49·994 mm is *high*.
> The *probability* of the machine producing a component the diameter of which is within the range 49·994 to 49·995 mm is *considerable*.
> The *probability* of the machine producing a component the diameter of which is within the range 49·995 to 49·996 mm is *low*.
> The *probability* of the machine producing a component the diameter of which is greater than 50·00 mm is *nil*.

Now complete these sentences:

> a) The probability of the machine producing a component the diameter of which is within the range 49·990 to 49·991 mm is _____ .
> b) The probability of the machine producing a component the diameter of which is within the range 49·988 to 49·987 mm is _____ .
> c) The probability of the machine producing a component the diameter of which is less than 49·980 mm is _____ .
> d) The probability of the machine producing a component the diameter of which is within the range 49·987 to 49·985 mm is _____ .
> e) The probability of the machine producing a component the diameter of which is within the range 49·990 to 49·988 mm is _____ .

7. Now look at this graph:

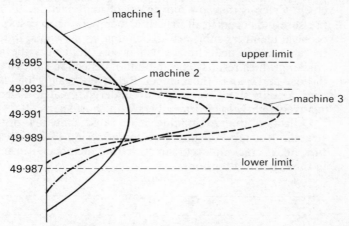

Component size distribution curve

Which machine is described in the following sentences?

> a) The probability of this machine producing a component greater than 49·995 mm in diameter is considerable.
> b) The probability of this machine producing a component greater than 49·995 mm is nil.

c) The probability of this machine producing a component the diameter of which is within the range 49·989 to 49·993 mm is very high.
d) The probability of this machine producing a component within the range 49·989 to 49·993 mm is relatively low.
e) The probability of this machine producing a component within the range 49·989 to 49·993 mm is relatively high.
f) The probability of this machine producing a component within the range 49·987 to 49·989 mm is low.

Section 3   Reading

8. Read this:

Every kind of crack which develops in a material is the result of that material failing to withstand the stresses which have been created in that part up to a given time. These cracks tend to result from the failure of the material to resist the tensile stresses created rather than
5   the compressive and shear stresses. From the structural point of view, the amount of reduction in strength caused by cracking will depend on the extent of the cracking which has occurred. It is for this reason that slight cracks are not usually considered as serious, as they do not result in a significant loss of strength. On the other
10   hand, it is important that a decision about whether cracks are serious or otherwise should be based on the relationship between the strength of the material and its shape and size relative to the stresses created.
    Cracking which results from external loads is usually associated with elastic deformations (deflection) and also occasionally with
15   creep. Creep is the slow movement of materials due to continuous pressure and occurs in all structural members. Most cases of cracking occur in members subjected to bending, as in beams. In these cases, the amount of deflection which occurs influences the amount of cracking which results. The amount of deflection varies according to
20   the load and span of the member. Relatively heavily loaded members with a long span are liable to undergo greater deflection than relatively lightly loaded members with short spans.
    Deflection due to loading may cause cracking of either the structural member or any cladding adhering to it such as plaster,
25   tiles etc. If cracking occurs in structural members, it usually occurs in those parts in tension.

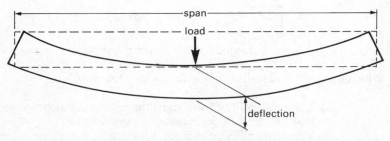

Deflection of a beam

The diagram illustrates a beam which has deflected under load. The bottom of the beam is in tension. When cracks develop on the lower surface there is likely to be a reduction of the structural area of the beam. The size of the cracks will influence the way in which stresses will be redistributed over the reduced area in order to resist the loads applied on the member.

30

Say whether the following statements are true or false. Correct the false statements.

a) All cracks are the result of tensile stresses.
b) The reduction in strength in a member which results from cracking depends on the size of the cracks.
c) Slight cracks are never considered to be serious.
d) Cracking is frequently associated with creep.
e) All structural members are liable to creep.
f) Cracking sometimes occurs in members subjected to bending stresses.
g) Greater deflection can be expected from members with long spans and a heavy load than from members with short spans and a light load.
h) Deflection may cause cracking in both structural and non-structural members.
i) Cracking always occurs in parts of a member subjected to tensile stresses.
j) There is a high probability that a member which has cracked under load will be less able to resist that load.

9. Answer these questions:

a) In line 8 what does 'this reason' refer to?
b) In line 17 what does 'in these cases' refer to?
c) In line 24 what does 'it' refer to?
d) In line 26 what does 'those parts' refer to?

10. Look again at the four paragraphs. Decide what each paragraph is about and give it a title. Then decide what the points made in the paragraph are, and put them under the title. The first one has been done for you:

The causes and effects of cracking
a) the relationship between stresses and cracks
b) factors which determine the importance of cracking

11. Complete the following:

When a part is subjected to stresses which it is not strong enough to withstand, ... Cracks, however, are more likely to be caused .... The amount of cracking which occurs ...... reduction in strength. Cracks are not considered serious provided .... However, the question of whether a crack is serious or not, is affected by ....

Along with cracking, ... also occur. Cracking depends on the amount of deflection which occurs, and the amount of deflection in turn depends on .... The greater the load on the member, and the longer the span, ...

Section 4 Listening

12. Below is a summary of the points in the listening passage, but they are in the wrong order. Put them in the correct order.

a) The disadvantages of water as a quenching agent.
b) The use of oil as a quenching agent.
c) The advantages of water as a quenching agent.
d) The advantages of brine over water.
e) The advantage of oil as a quenching agent.
f) The effect of steam pockets, and how the pockets can be removed.

13. Answer these questions:

a) Name two disadvantages of water in quenching.
b) What is the advantage of brine in quenching?
c) What is the advantage of oil in quenching?
d) Why is water excellent for the quenching of small parts?

Unit 11   Process 4 Method

Section 1   Presentation

1. Look and read:

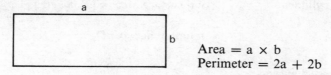

Area = a × b
Perimeter = 2a + 2b

A rectangle (all corners are right angles)

A rectangle *is defined as* a quadrilateral with opposite sides equal and parallel and whose corners are right angles.

The area of a rectangle $\left\{ \begin{array}{c} is \\ can\ be \end{array} \right\}$ calculated by multiplying its length by its width.

The perimeter $\left\{ \begin{array}{c} is \\ can\ be \end{array} \right\}$ calculated by *adding together* the length of all the sides i.e. length + length + width + width.

Now make similar statements about the following:

a)

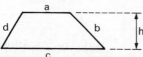

Area = a × h
(where: h = the perpendicular
height)
Perimeter = 2a + 2b

A parallelogram
(opposite sides parallel, corners not at right angles)

b)

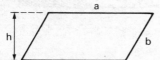

Area = $\dfrac{a + c}{2}$ × h

(where: $\dfrac{a + c}{2}$ = the average length)
Perimeter = a + b + c + d

A trapezium
(two sides parallel, corners not at right angles)

c)

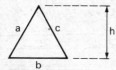

Area = ½base × height
    = ½(b × h)
Perimeter = a + b + c

An equilateral triangle
(all sides equal)

d)

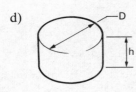

A solid cylinder

Cross-sectional area $= \pi\dfrac{D^2}{4}$

Volume = cross-sectional area

$\times$ height $= \left(\pi\dfrac{D^2}{4}\right) \times h$

Total surface area $= \left(2\ \pi\dfrac{D^2}{N}\right) + \pi Dh$

Circumference $= \pi D$

2. Look at these diagrams:

(i) The steel rule

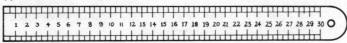

(ii) The vernier caliper

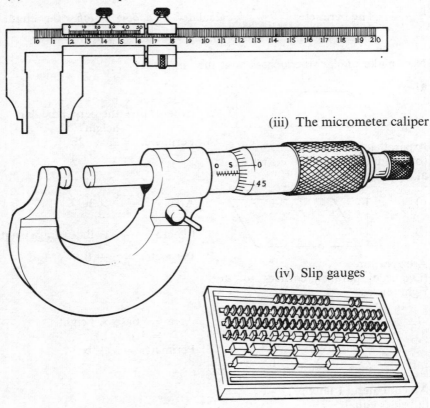

(iii) The micrometer caliper

(iv) Slip gauges

Now look at this table which describes the accuracy and applications of measuring instruments:

Name	Reading accuracy (mm)	Typical application
Steel rule	0·5	measuring the distance between lines. measuring the distance between faces.
Vernier caliper	0·02	accurate measurement of internal and external diameters of ring-shaped components.
Micrometer caliper	0·01	accurate measurement of diameters (external only) and thicknesses. Quicker and easier to use than the Vernier, but limited range for any one instrument.
Slip gauges	0·0025	a workshop standard used for checking and setting measuring instruments and workshop gauges.

Ask and answer questions like these examples:

Question: Which instrument is used to measure the distance between lines?
Answer: A steel rule.
Question: Which instruments are more accurate than the vernier caliper?
Answer: The micrometer caliper and the slip gauges.
Question: If we want to measure the internal diameter of a component, which instrument should we use?
Answer: We should use a vernier caliper.

3. Look at these diagrams:

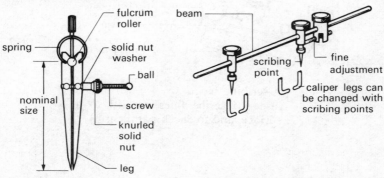

Dividers
(used to scribe circular lines up to about 150 mm in diameter)

Trammels
(used to scribe circular lines above 150 mm in diameter)

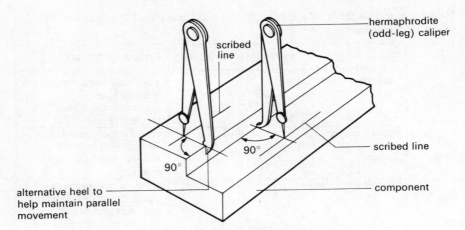

A hermaphrodite (odd-leg) caliper
(used to scribe lines parallel to an
edge)

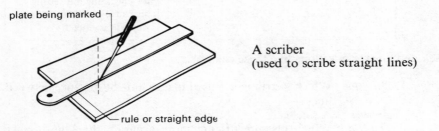

A scriber
(used to scribe straight lines)

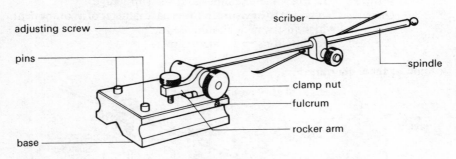

A scribing block
(used to scribe lines parallel to a
surface, and to check parallelism)

Make true sentences from this table:

To In order to	scribe	circular lines up to about 150 mm in diameter lines parallel to a surface straight lines circular lines above 150 mm in diameter lines parallel to an edge

dividers a scriber a hermaphrodite caliper a scribing block trammels	is are	normally used.

Section 2   Development

**4. Read these instructions. They are the instructions for carrying out a
tensile test on a material:**

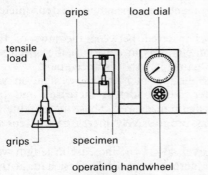

grips   load dial

tensile
load

grips ─   specimen

operating handwheel

Tensile testing machine

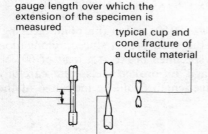

gauge length over which the
extension of the specimen is
measured

typical cup and
cone fracture of
a ductile material

thinning down or 'necking' of a
ductile specimen immediately before
fracture

Tensile test specimen

Insert the specimen into the jaws of
the testing machine.

Attach an extensometer to the specimen with its locating points set at
gauge length distance apart.

Apply a tensile load to the specimen
and measure the elongation on the
extensometer.

Increase the load in steps of 100 kN
up to the elastic limit.

At the elastic limit remove the
extensometer, and reload the specimen till fracture takes place.

Remove the specimen, and fit together the broken ends in order to
measure the gauge length at the
moment of fracture.

Now rewrite these instructions in the same way as this example:

First of all the specimen *should be inserted* in the jaws of the testing machine.

Note: where there are two verbs, we do not need to repeat *should* e.g. A tensile load *should be applied* and the elongation *measured* on the extensometer.

5. Read this:

A laboratory report normally gives the *object* of the experiment, the *equipment* which was used, the *procedure* which was followed, the *results* which were obtained and the *conclusions*.

Read the following, which is a report on a tensile test carried out on an annealed mild steel specimen.

Tensile test

Object: To determine the load/elongation curve for the mild steel specimen provided. To calculate for the specimen the: (i) tensile strength (ii) percentage elongation (indication of ductility).

Equipment: Tensile testing machine, specimen, extensometer, dividers, rule.

Procedure: The specimen was inserted between the jaws of the testing machine. The extensometer was attached to the specimen with its locating points set at 'gauge length' distance apart.

A tensile load was applied to the specimen and the elongation was measured on the extensometer. The load was increased and the corresponding elongation was again measured. These measurements were repeated as the load was progressively increased in steps of 10 kN.

It was noticed that at just over 40 kN the increase in length was no longer proportional to the increase in load. The test had, at this point, proceeded beyond the elastic limit of the specimen ('E' on the graph). The extensometer was removed so that it would not be damaged by the specimen fracturing.

Subsequent measurements were taken using dividers and a steel rule.

The maximum load applied was 80 kN and at this point 'necking' commenced. Then although the load was reduced, elongation continued until eventually the specimen broke at a load of 60 kN. The specimen was removed, and its broken ends were carefully fitted together so that the gauge length could be measured at the moment of fracture.

Results:

Load (kN)	0	10	20	30	40	47	45	50	60	70	80	70
Elongation (mm)	0	0·125	0·188	0·25	0·375	1·0	1·5	2·1	3·9	6·0	11·5	14·5

Load at moment of fracture 60 kN; elongation 16·5 mm

Specimen: Annealed mild steel
diameter 15 mm
gauge length 75 mm

Graph:

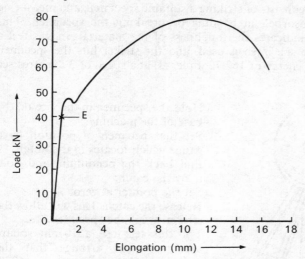

Load/elongation curve for mild steel

Calculations:

Tensile strength $= \dfrac{\text{maximum load}}{\text{original cross-section area}}$

$= \dfrac{80{,}000\,\text{N}}{3{\cdot}142 \times (7{\cdot}5)^2}$

$= 453\,\text{N/mm}^2$

Now calculate the percentage elongation:

Percentage elongation $\% = \dfrac{\text{elongation at fracture} \times 100}{\text{original gauge length}}$

Now complete the conclusion:

Conclusion: The specimen showed ____ properties up to a load of approximately 40 kN. The maximum load was approximately ____, after which ____ commenced. At this point the stress increased up to the point at which the specimen ____. This occurred at a load of approximately ____. The calculations showed that the specimen had a percentage elongation of ____. This indicates that mild steel has a relatively high ....

6. Here is a set of instructions for an Izod test, which is a test of the toughness of materials. Imagine you have carried out this test and obtained

99

the results given after the instructions. Write a full laboratory report with the object of the test, the equipment used, the procedure and results of the test. (There is no graph with this experiment.)

The Izod test consists of striking a suitable specimen, and measuring the energy absorbed in bending or breaking the specimen. This energy value indicates the toughness of the material. In this test a 10 mm square specimen is used, and the striker hits the specimen with a kinetic energy of 162·72 joules (J) at a speed of 3·8 metres/sec (m/s).

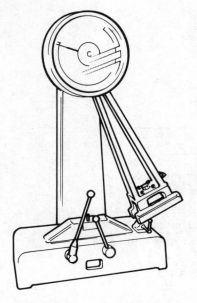

Hold the specimen in the vice on the base of the machine.
Set the specimen in position by a gauge which locates in the notch.
Pull back the pendulum, and hold it by the catch.
Set the pointer at zero.
Release the catch. This will allow the striker to hit the specimen.
For this test use a 10 mm square specimen, and arrange that the striker hits the specimen with 162·72 joules (J) kinetic energy at 3·8 metres/sec (m/s).

Typical values
Stainless steel	150J
Manganese steel	70J
High duty cast iron	14J
Aluminium alloy (LM 5)	8J
An alloy steel containing chromium nickel and molybdenum (AMS6407)	22-2J

Now write a conclusion comparing the results obtained. Use the following phrases:

The maximum value obtained was ____ .
The minimum ....
The toughest material ....
The least tough ....
____ has a relatively high value, while ____ has a relatively low value.
The probability of ____ fracturing on impact is high/low etc.

Section 3   Reading

7. Read this:

One method of determining the workability of concrete on site, where accurate measurements cannot be taken, is to carry out the slump test. By means of this test the consistency i.e. the ease of flow of the concrete can be measured; this depends on the proportion of water to cement and the size of the aggregate.

The equipment needed consists of a metal container in the shape of a truncated cone, a flat plate, a 16 mm diameter rod and a rule. The container has a diameter of 100 mm at the closed end, 200 mm at the other end and a height of 300 mm.

Freshly mixed concrete made with aggregate of less than 50 mm diameter is put into the cone in three layers. Each layer is compacted 25 times with the rod before the next layer is added. The cone is inverted over the plate and removed. The amount that the concrete settles or slumps shows its consistency.

This test is often used on building sites to give an approximate measure of consistency. It can be done quickly, it is easy to teach and it can be done without expensive equipment.

Chart of the max/min permitted slump for different
structural purposes

| Portion of structure | Consistency slump | |
	Maximum (in mm)	Minimum
Reinforced foundation walls and footings	125	50
Building columns	150	75
Pavements	75	50

Now answer these questions:

a) What is the purpose of the slump test?
b) Is the slump test the only method of determining the rate of flow of concrete?
c) Does the slump test give an accurate measurement of workability?
d) How is the workability of the concrete calculated?
e) What are the advantages of the slump test?
f) What can concrete which slumps 150 mm be used for?
g) If the concrete slumps 100 mm, can it be used for pavements?
h) If the concrete slumps 50 mm, can it be used for building columns?

8. Continue reading:

Another test which measures the consistency of concrete is known as the compacting factor test. It is more accurate than the slump

101

test, but it is more difficult to carry out and requires more accurate measurement.

The apparatus consists of three cylindrical containers placed one above the other, supported by an iron frame. The upper two containers can be opened at the bottom to allow the concrete to fall into the third. A vibrator for compacting the concrete and an accurate set of scales for weighing the concrete are necessary.

To carry out the test, the lowest container is completely filled with compacted concrete and weighed. This concrete is then emptied and the upper container is filled with the sample to be tested. The bottom is opened, thus allowing the concrete to fall into the middle container which is too small to contain the total volume. The bottom of the middle container is then opened and the falling concrete more than fills the bottom container. The excess concrete is removed so that the surface is level with the top of the container. The full container is then weighed. This weight is divided by the weight of the fully compacted concrete.

The compacting factor test gives more accurate information than the slump test, but, as it needs slightly more elaborate equipment, it is more often used in the laboratory than on site.

Now write instructions for a building worker to carry out the compacting factor test, using the following words:

first, then, after this, finally

Example: First of all, fill the lowest container with compacted concrete.

9. Look at the following outline of the first reading passage:

The slump test
a) the purpose of the slump test
b) the equipment used in the slump test
c) the method of carrying out the slump test
d) the uses and advantages of the slump test

Now make a similar outline of the second reading passage.

10. Make a list of the differences between the slump test and the compacting factor test, under these headings:

Slump test	Compacting factor test
a) Approximate measurements	Relatively accurate measurements

11. Look at these sentences:

One difference between the slump test and the compacting factor

test is that $\left\{ \begin{array}{l} \textit{the former} \\ \textit{the first} \end{array} \right\}$ gives approximate measurements $\left\{ \begin{array}{l} \textit{whereas} \\ \textit{while} \end{array} \right\}$

$\left\{ \begin{array}{l} \textit{the latter} \\ \textit{the second} \end{array} \right\}$ gives relatively accurate measurements.

$\left\{ \begin{array}{l} \textit{They} \\ \textit{The tests} \end{array} \right\}$ differ in that the slump test gives approximate measurements

$\left\{ \begin{array}{l} \textit{whereas} \\ \textit{while} \end{array} \right\}$ the compacting factor test gives relatively accurate measurements.

Now make similar statements about the differences you have listed.

Section 4   Listening

12. Listen to the passage, and make notes on the following points as you listen:

    a) the purpose of the crushing test
    b) the normal size of the specimen in Britain
    c) how the specimen is kept ready for testing after it has been removed from the mould
    d) the reason for curing the concrete
    e) when the specimen is normally tested
    f) how the crushing strength is calculated

13. Now answer these questions:

    a) What is the purpose of the crushing test?
    b) What size of specimen is usually tested in Britain?
    c) Where is the specimen kept after it has been removed from the mould?
    d) Why is the specimen cured?
    e) How many days after being mixed is the concrete tested?
    f) How is the crushing strength calculated?

14. Now listen to the passage again, and then write instructions for carrying out the crushing test.

Unit 12   Consolidation

1. Look and read:

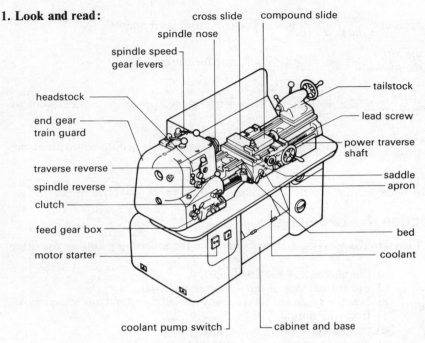

A centre lathe

A lathe is a machine tool which is used to produce work which is circular in cross-section by rotating the work against a cutting tool. In other words, it generates a surface of revolution. The finished work may be cylindrical or tapered (i.e. cone-shaped). In addition to generating cylindrical surfaces, lathes can generate plane surfaces by facing. The lathe may also be used for boring and cutting screw threads.

Indicate which sentences and words express the following:

    a) a definition
    b) three-dimensional shapes
    c) a two-dimensional shape
    d) a possibility
    e) three operations carried out on components
    f) statements of function
    g) a statement of ability

2. Continue reading:

One type of lathe is known as a centre lathe, and it is made up of a number of basic parts which are accurately aligned one to another so that accurate components can be produced.

The bed acts as a support for the other parts and, therefore, consists of a strong, rigid framework usually made of cast iron. The top is machined to form the slideways, which carry the saddle, the headstock and the tailstock. The saddle controls the movement of the cutting tool. It is free to slide backwards and forwards along the slideways, thus enabling the tool to move parallel to the spindle axis. On the upper surface of the saddle is the cross-slide, which enables the tool to move at right angles to the axis of the spindle, and is, therefore, used in facing operations.

The compound slide is mounted on the upper surface of the cross slide. This can be rotated in such a way that the tool is able to move at an angle to the spindle axis when cutting a taper. The tool post is mounted on the compound slide, and carries the cutting tool.

The other principal parts on the lathe are the headstock and the tailstock. The headstock contains the gears, the controls and the spindle to which the workpiece is attached. It is the spindle which causes the workpiece to rotate. The tailstock does not rotate and is used only to support the other end of the workpiece.

Complete the following:

The second part of the reading passage is concerned with:
a) ...... of one kind of lathe by its composition.
b) ...... of various parts of the lathe, and description of their ......
and ....

3. Look at the diagrams and answer the questions:

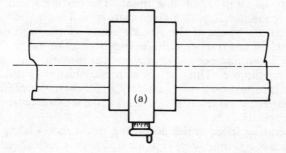

Basic lathe structure

Which part of the lathe is (a)?
Indicate the directions of its movements.

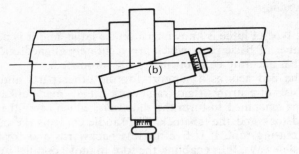

Basic lathe structure

Which part of the lathe is (b)?
Indicate the directions of its movements.

Now complete this table:

Part	Location	Function
Headstock Tailstock Saddle Tool post Cross slide Compound slide Bed		

4. Continue reading:

The gears in the headstock enable the operator to change the speed
at which the workpiece is rotated. The speed chosen for a given
material depends on the size of the diameter of the workpiece and
the material from which it is made. The cutting speed is constant
5   for a given material, with a given cutting tool for a given job.

The cutting speed is the surface speed of the workpiece and is
measured in metres per minute and is determined by the spindle
speed, i.e. the number of revolutions per minute, and the diameter
of the workpiece. Thus at constant revolutions surface speed is
10   directly proportional to diameter. It follows, therefore, that the
spindle speed is inversely proportional to the diameter for a constant
surface speed.

The cutting speed varies according to the metal being machined.
For example, if mild steel is machined with a high speed steel tool,
15   the cutting speed should be 18 to 24 metres per minute. If, however,
aluminium is machined with the same cutting tool, the cutting speed
should be 92 to 150 metres per minute. These speeds are for rough
cuts; for finishing cuts the speed may be increased by approximately
15%.

Now say whether these statements are true or false. Correct the false statements.

a) The cutting speed depends only on the diameter of the workpiece.
b) Spindle speed is measured in rev/min.
c) The greater the diameter of the workpiece, the lower the spindle speed.
d) For a given spindle speed we can say that the greater the diameter of the workpiece, the lower the cutting speed.
e) The cutting speeds for roughing cuts are normally lower than for finishing cuts.
f) In line 18 'speed' refers to spindle speed.

5. Complete the following outline of the reading passage above:

The speed of operation of the lathe
The function of the ____
The factors determining ____
Speed in relation to the work being done, the ____ and the ____

Cutting speed
____ and ____ of cutting speed
                    The factors ......
The relationship between ____ and ____
The relationship between ____ and ____

Metals and cutting speed
The relationship between ......
Two ____ of the relationship ......
The difference between ......

6. Continue reading:

A chuck is frequently used to hold the workpiece to the lathe. It is fixed to the rotating spindle and consists of a circular body with four jaws, which are adjusted by means of screws. The jaws can be moved inwards and outwards so that the diameter of the circle
5   formed by the jaws can be increased or decreased.
The workpiece which is to be worked upon is inserted into the space between the jaws. The screws are then tightened, with the result that the workpiece is gripped tightly. It is then necessary to test whether the centre of the workpiece is in alignment with the
10   centre of the spindle. This can be done by rotating the workpiece slowly against a piece of chalk held in a fixed position. If the chalk draws an even line, the workpiece is centred. If the line is uneven or broken, the jaws must be adjusted.
In case of irregular workpieces, the weight on one side of the
15   chuck must be balanced by an extra weight on the other side, so that the forces on the spindle are in equilibrium. If the forces are not in equilibrium, the resulting vibration will lead to a poor surface.

Now answer these questions:

- a) What does 'it' in line 1 refer to?
- b) What does 'this' in line 10 refer to?
- c) What is adjusted by means of the screws?
- d) What is the function of the jaws?
- e) How is the alignment of the centre of the workpiece with the centre of the spindle tested?
- f) Does the passage tell us why this alignment is tested?
- g) Does vibration occur with both regular and irregular workpieces? What causes vibration?

Now make an outline of the passage (6) similar to the one you completed after reading passage (4).

7. Look and read:

The tensile test

The tensile strength and ductility of a material are determined by carrying out a tensile test. A standard test piece is fixed between two grips which are fitted to the testing machine. A load is applied to the test piece and the extension is recorded. Most modern testing machines have recorders which automatically produce a load-extension graph during the test. On the graph for mild steel four distinct points can be marked off. These are as follows:

The elastic limit:

This is the point up to which a material remains elastic. In other words, it is the point beyond which a material ceases to return to its original shape or size on removal of the load to which it is subjected. After the material takes on permanent set, i.e. it will not return to its original size, it is said to be plastic.

The yield point:

This is the point at which the material suddenly extends without an increase in load.

The maximum load:

This is the point at which the applied load reaches its maximum. After this point the phenomenon known as 'necking' or 'waisting' is observed. This is the phenomenon in which the specimen becomes much thinner prior to breaking.

The breaking point:

This is the actual point at which the specimen breaks. The load at this point is less than that at the maximum load.

Copy this graph, and put in the four points described in the passage (7):

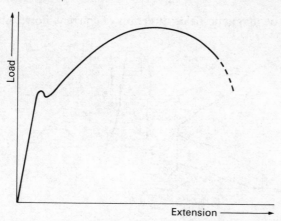

Load/extension curve for annealed low carbon (dead mild) steel

Now answer these questions:

    a) What is the purpose of carrying out a tensile test?
    b) Explain the difference between a plastic and an elastic material.
    c) Define 'necking' and 'permanent set'.
    d) What is the difference between the elastic limit and the yield point?
    e) Does the writer say that this difference between the elastic limit and the yield point is true for all materials?
    f) Does the specimen in a tensile test break at a load greater or smaller than the maximum load?

8. Now read this:

Cutting of magnetic lines of force

The cutting of the magnetic flux by a conductor is the most important method of producing electricity and is the only way in which mechanical energy can be converted into electrical energy. When a conductor cuts a magnetic field at right angles to the field, or vice versa, an electrical current will flow in a conductor if it is connected into an external circuit. Fleming's right hand rule states that if the thumb, forefinger and second finger of the right hand are held at right angles to each other, and if the thumb indicates the direction of movement of the conductor, then the forefinger will indicate the direction of the magnetic field, and the second finger will indicate the direction of current flow.

Now match the letters in the diagram with these labels:

direction of magnetic field, direction of current flow, direction of motion

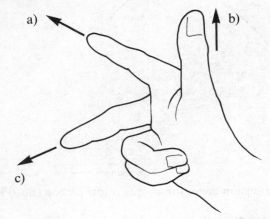

Fleming's right-hand rule for generators

Look at the diagram below. Let the direction of the movement of the conductor be upwards. With the help of Fleming's right hand rule, work out the direction of the magnetic field, and the direction of the current flow.

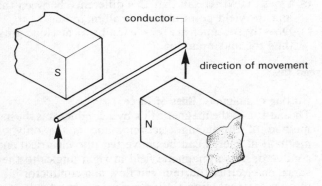

Induction in a conductor

9. Continue reading:

Machines that produce electricity in this way are called generators. A simple generator consists of a coil of wire which is free to rotate between the opposite poles of a U-shaped magnet. The ends of the coils are brought out to metal collector rings, which slide over the ends of stationary carbon brushes. The current is led away by these brushes. The coil is rotated by mechanical means.

Maximum current is generated when the coil sides cut the magnetic field at right angles. When the coil sides move parallel to the magnetic field, no lines of force are being cut, and, therefore, no current is generated. In between these positions, the current varies between zero and maximum, the value depending on the angle at which the coil cuts the magnetic field.

Now answer these questions:

a) What does 'in this way' in line 1 refer to?
b) Which parts of the generator in this example are fixed and which are free to rotate?
c) How is mechanical energy converted into electrical energy in a simple generator?
d) Is the current produced by a generator constant?
e) What must we know if we want to calculate the current at a particular time?

Now draw diagrams of the positions of the coil to show how the current varies between maximum and minimum in a simple generator.

Glossary

This list gives the pronunciations of the technical and semi-technical words used in this book and definitions of those words that are not fully explained in the text or diagrams. An asterisk (*) means that a word in the definition is itself explained in the Glossary. The number after each entry indicates the unit in which the word first appears.

Pronunciations are shown in the system that is used in the new Longman *Dictionary of Contemporary English*. The symbols are shown in this table, with a key word for each. The letters printed in **bold type** represent the sound value of the symbol.

Consonants

p	pea	f	few	ʃ	fishing	h	hot
b	bay	v	view	ʒ	pleasure	m	sum
t	tea	θ	thing	tʃ	choose	n	sun
d	day	ð	then	dʒ	jump	ŋ	sung
k	key	s	soon	l	led	j	yet
g	gay	z	zoo	r	red	w	wet

Vowels

iː	sheep	ɔː	caught	eɪ	make	ɪə	here
ɪ	ship	ʊ	put	əʊ	note	eə	there
e	bed	uː	boot	aɪ	bite	ʊə	poor
æ	bad	ʌ	cut	aʊ	now	eɪə	player
ɑː	calm	ɜː	bird	ɔɪ	boy	əʊə	lower
ɒ	cot	ə	about			aɪə	tire
						aʊə	tower
						ɔɪə	employer

Notes

1. A small raised /ʳ/ at the end of a word means that the /r/ is pronounced if a vowel follows (at the beginning of the next word), but not otherwise. For example, *far* /fɑːʳ/ means that *far away* is pronounced /fɑːr əweɪ/ but *far down* is /fɑː daʊn/.

2. The italic /ə/ means that the sound /ə/ can be used but is often omitted. It may be found before the consonants /m, n, ŋ, l, r/ in certain positions. For example, *travel* /'trævəl/ means that the pronunciation /'trævəl/ is possible but /'trævl/ may be more common.

3. The mark /'/ means that the following syllable has *main stress*, and /,/ means that the following syllable has *secondary stress*. For example, *understand* /,ʌndə'stænd/.

113

abrasion /əˈbreɪʒn/ rubbing, wearing away 8

adhering /ædˈhɪərɪŋ/ sticking fast to 10

adiabatic expansion, contraction /ˌeɪdaɪə ˌbætɪk ɪkˈspænʃən, kənˈtrækʃən/ expansion or contraction in which there is no passing in or out of heat 6

alternating current /ˌɔːltəneɪtɪŋ ˈkʌrənt/ an electric current which first increases to a maximum in one direction, decreases to zero and then increases in the opposite direction and so on 5

anchor /ˈæŋkəʳ/ a device used to secure a ship 1

applications /æplɪˈkeɪʃənz/ uses 10

bar /bɑːʳ/ a material of uniform cross-section 1

basic allowable stress /ˌbeɪsɪk əˌlaʊəbəl ˈstres/ the stress which is considered to be a safe maximum in normal conditions 7

battery terminal /ˈbætrɪ, tɜːmɪnəl/ screw into which the ends of the wires are fixed 1

beam /biːm/ a structural member designed to resist bending loads A

blast of air /ˌblɑːst əv ˈeəʳ/ a jet of compressed air

boring /ˈbɔːrɪŋ/ cutting holes 12

bulldozer /ˈbʊldəʊzəʳ/ a machine with a large plate in front used to push masses of earth 8

car body panel /ˈkɑː bɒdɪ, pænəl/ a sheet of metal which has been made into a car body component e.g. door, wing 3

carburettor /kɑːbəˈretəʳ/ instrument which mixes air and petrol to make an explosive mixture 8

car radiator /ˌkɑː ˈreɪdieɪtəʳ/ the apparatus which cools a car engine 1

cast /kɑːst/ to melt (metal) and pour it into a mould.

cast iron /ˌkɑːst ˈaɪən/ a brittle form of iron containing more than 1·7%, and usually 2·4–4% carbon 1

centre /sentə/ to place in the exact centre for drilling 5

ceramics /səˈræmɪks/ 3

chain /tʃeɪn/ a number of links joined together to form a line 1

charge /tʃɑːdʒ/ to place limestone, coke etc. in a blast furnace 6

chisel /ˈʃɪzəl/ a tool for shaping wood, stone or metal 1

cladding /ˈklædɪŋ/ covering 10

clay /kleɪ/ a material which becomes

hard when baked. Bricks and pots are made from it 6

clinker /ˈklɪŋkəʳ/ incombustible residue from furnaces used for road-making and as aggregate for concrete 6

coarse /kɔːs/ having large crystals 3

coat /kəʊt/ to cover 5

cold work /ˈkəʊld wɜːk/ to work metal without it being heated 3

column /ˈkɒləm/ a vertical support 7

concentrate /ˈkɒnsəntreɪt/ to reduce the amount of earth matter (gangue) in an ore 6

condenser /kənˈdensəʳ/ an instrument for cooling steam and changing it back to water 6

converter (kənˈvɜːtəʳ/ a furnace used in the process of steel making 6

crane /kreɪn/ a machine for lifting and carrying heavy weights 1

crankshaft /ˈkræŋkʃɑːft/ a device for changing a linear movement into a circular movement, or a circular movement into a linear one 5

critical point /ˌkrɪtɪkl ˈpɔɪnt/ in heat treatment, the temperature at which a change in the molecular structure of a carbon steel occurs C

crush /krʌʃ/ to break into small pieces by pressure 6

data /ˈdeɪtə/ facts 7

deform /dɪˈfɔːm/ to change the shape of a body 1

dent /dent/ to leave an impression on a surface 1

displacement /dɪsˈpleɪsmənt/ the difference or geometrical relationship between the first position of a body and a later position 4

draw /drɔː/ to pull 1

ductile /ˈdʌktaɪl/ 1

ductility /dʌkˈtɪlətɪ/ 1

elasticity /ɪlæˈstɪsətɪ/ quality which makes a thing return to its first shape after the removal of load 1

electrolysis /ɪlekˈtrɒləsɪs/ the conduction of electric current through an electrolyte 6

electrolyte /ɪˈlektrəlaɪt/ a liquid which conducts electricity 6

e.m.f. /ˌiːemˈef/ electromotive force, i.e. the force which makes electric current flow in a circuit 3

energise /ˈenədʒaɪz/ to put energy into e.g. an electrical circuit 5

erect /ɪˈrekt/ to build 7

exhibit /ɪg'zɪbɪt/ to show 1

extensometer /eksten'sɒmɪtəʳ/ an instrument used for measuring the increase in length when a metal bar is stretched 11

failure /'feɪljəʳ/ the condition at which a structure cannot carry out its function, because of cracking, deformation etc. 8

failure rate /'feɪljə reɪt/ the percentage number of failures 10

fatigue /fə'tiːg/ the lowering of the breaking load of a member by repeated changes of stress 8

feed pump /'fiːd pʌmp/ the pump which pumps the water supply into a boiler at boiler pressure 6

filler metal /'fɪlə metl/ the metal added to join two parts in welding 5

fine /faɪn/ to have small crystals 3

fire /faɪə/ to bake 6

fire brick /'faɪə brɪk/ a brick with high resistance to heat used in furnaces 1

fluid /fluːɪd/ a substance which flows, i.e. a gas or a liquid 6

fluidity /fluːɪdətɪ/ the ability to flow 1

footing /'fʊtɪŋ/ the widening at the bottom of a wall 11

forging /'fɔːdʒɪŋ/ a piece of metal that has been shaped by a press or hammer C

foundation /faʊn'deɪʃn/ the strong base of a building or road 11

galvanised steel /ˌgælvənaɪzd 'stiːl/ steel coated with zinc which therefore has higher resistance to rust 5

granular /'grænjʊləʳ/ made up of small grains 3

gravity /'grævətɪ/ the force which attracts bodies to the centre of the earth C

grind /graɪnd/ to cut by abrasion* C

grip /grɪp/ to hold tightly 11

gypsum /'dʒɪpsəm/ crystalline calcium sulphate 6

hammer /'hæməʳ/ to hit with a hammer 1

heat capacity /'hiːt kə,pæsətɪ/ the ability to retain heat 5

heat exchanger /'hiːtɪks,tʃeɪndʒəʳ/ an apparatus in which heat is transferred from one material to another e.g. from a hot gas into steam 5

high speed steel /ˌhaɪ spiːd 'stiːl/ a hard steel used for metal cutting tools. It usually contains 12 to 18% tungsten, up to 5% chromium 0·4 to 0·7% carbon, and small amounts of other elements 10

hook /hʊk/ a curved piece of metal for catching hold of an object 5

horseshoe magnet /ˌhɔːsʃuː 'mægnɪt/ a magnet shaped like a horseshoe 5

house /'haʊz/ to contain within 5

hydraulic cement /haɪ,drɔːlik sɪ'ment/ a cement which requires the addition of water to make it harden 6

impurities /ɪm'pjʊərətɪz/ substances which are harmful to the material in which they are found

induce /ɪn'djuːs/ (as in induced current) to produce current 9

intermittent /ɪntə'mɪtənt/ not operating continuously 8

isothermal expansion, contraction /aɪsə,θɜːməl ɪk'spænʃn, kən'trækʃn/ the change which takes place without any change in temperature 6

jaw /dʒɔː/ parts of a machine between which objects are gripped 11

kinetic energy /kɪ,netɪk 'enədʒɪ/ the energy of a moving body 4

lightweight /'laɪtweɪt/ low mass* 5

limestone /'laɪmstəʊn/ rock formed from calcium carbonate deposits 6

load /ləʊd/ the force applied to a part 1

locating points /ləʊ'keɪtɪŋ pɔɪnts/ the points used to locate the specimen in a tensile test in the jaws of the testing machine 11

longitudinally /lɒŋgɪ'tjuːdɪnəlɪ/ 2

machine /mə'ʃiːn/ to cut a metal with a machine 3

magnetic flux /mæg,netɪk 'flʌks/ the total amount of magnetic power 9

mass /mæs/ the quantity of matter that a body contains 1

mechanical link /mə,kænɪkl 'lɪŋk/ solid connection between two points of a mechanism 6

moist /mɔɪst/ slightly wet 1

momentum /mə'mentəm/ energy stored in a moving body 3

Monel metal /mɒ'nel/ an alloy of nickel and copper 4

operator /'ɒpəreɪtəʳ/ a person who operates or works a machine 5

optimum /'ɒptɪməm/ best 9

pavement /'peɪvmənt/ in Britain, the footway at the side of a road 11

periphery /pə'rɪfərɪ/ outside edge 8

pig iron /'pɪg aɪən/ iron as first produced from the ore 6

pine /paɪn/ wood of the pine tree 7

plasticity /plæs'tɪsətɪ/ the quality which makes a thing keep its changed shape after the removal of load   1

polarisation /pəʊləraɪzeɪʃn/ the gathering of a chemical substance on one of the plates of an electric cell with the result that the cell loses some of its power   B

potential energy /pə,tentʃəl 'enədʒɪ/ energy which may be used because of its position   4

powder /'paʊdəʳ/ to make into powder   6

quenching /'kwentʃɪŋ/ the cooling of hot metal in water or oil   C

raw material /,rɔː mə'tɪərɪəl/ material in its natural state i.e. ore, coal   6

reduce /rɪ'djuːs/ to take oxygen from   6

reinforcement /,riːɪn'fɔːsmənt/ strengthening (for concrete)   A

restrain /rɪ'streɪn/ to hold in place   2

roast /rəʊst/ to heat sulphide ores in air to remove the sulphur and to change them to oxides   6

rod /rɒd/ a straight piece of metal which has a small cross section compared with its length   1

rolling /'rəʊlɪŋ/ a method of reducing the section of a bar   A

rusty /'rʌstɪ/ affected by rust i.e. the brownish coating formed on iron or steel by oxidation   1

scratch /skrætʃ/ to make lines on a surface with something sharp   1

sheath /ʃiːθ/ a cover   5

size /saɪz/ to arrange according to size   6

slag /slæg/ waste material and impurities which come to the top of melted metal   6

slideways /'slaɪdweɪz/ the part on which the sliding part of a machine runs   12

solder /'səʊldəʳ/ lead and tin (or other easily melted alloy) used to join metals together   4

span /spæn/ distance between two ends   10

spindle /'spɪndl/ a machine element acting as the revolving axis on which another element revolves. The cylindrical member revolving with the headstock of a lathe   2

stainless steel /,steɪnləs 'stiːl/ an alloy of steel with chromium or chromium and nickel which is not affected by water or weak acid   3

suspension bridge /sə,spenʃn ,brɪdʒ/ bridge held up from above by rods and chains   1

tear /teə/ to come apart   C

tonne /tɒn/ 1,000 kilograms   7

truncated cone /trʌŋ,keɪtɪd 'kəʊn/ cone with the end cut off   10

turbine /'tɜːbaɪn/ an engine in which the rotary motion is produced directly by the power of a jet of steam or water acting on blades set round a shaft   3

tuyere /'twiːjə/ pipe through which hot air is blown into a furnace   6

uniformly /juːnɪ 'fɔːmlɪ/ without varying   10

vacuum /'vækjuːm/ a space completely empty of substance or gas   8

valve /vælv/ a device which allows gases or liquids to flow in one direction only (in or out)   3

vibration /vaɪ'breɪʃn/ rapid movement backwards and forwards   12

weldable /'weldəbl/ it can be welded   A

withstand /wɪθ'stænd/ to resist   7

workability /wɜːkə'bɪlətɪ/ (concrete) the ease with which concrete can be placed   11

wrought iron /,rɔːt 'aɪən/ a very malleable iron with a very low carbon content and containing some slag   3

116

Bibliography

BENNET, G. A. G. *Electricity and Modern Physics*. Edward Arnold, 2nd edn, 1974.

GREER, A. AND HOWELL, W. H. *Mechanical Engineering Craft Studies Part 2*. Edward Arnold, 1972.

MC KAY, W. B. *Building Construction*. Longman, 1971.

SCOTT, J. S. *A Dictionary of Civil Engineering*. Penguin Books, 1958.

TIMINGS, R. L., WATSON, W. G. Longman Craft Studies Series. *Basic Electrical and Electronic Engineering*. Longman, 1973.

TIMINGS, R. L. Longman Craft Studies Series. *Basic Engineering*. Longman, 1973.

TIMINGS, R. L. Longman Craft Studies Series. *Basic Mechanical Engineering*. Longman, 1974.

TIMINGS, R. L. Longman Craft Studies Series. *Mechanical Engineering*. Longman, 1972.

TIMINGS, R. L. Longman Technician Series. *Processes and Materials*. Longman, 1978.